COSMIC NOMAD

JOSH RICHARDS

First edition - published in 2020 by Launchpad Speaking
Copyright © Josh Richards, 2020
The moral rights of the author have been asserted

All rights reserved. Except as permitted under the Australian Copyright Act 1968 (for example, a fair dealing for the purposes of study, research, criticism or review), no part of this book may be reproduced, stored in a retrieval system, communicated or transmitted in any form or by any means without prior written permission.

All inquiries should be made to the author.
josh@joshrichards.space
www.joshrichards.space

Printed in Australia by Ingram Spark
ISBN (Paperback Edition) 978-0-6481356-2-3
1. Wit and Humour 2. Science

Cover design by Anna Piper Scott
Foreword by Dr Niamh Shaw
Editing by Carly Bodnar, Georgi McLaren, Chloe Reid, Shelley Richards, and Lisa Stojanovski

Discounts are available on bulk orders by corporations, associations, and others. Contact the author at the email above for details.

Disclaimer: The material in this publication is of the nature of general comment only, and does not represent professional advice. It is not intended to provide specific guidance for particular circumstances and it should not be relied on as the basis for any decision to take action or not take action on any matter which it covers. Readers should obtain professional advice where appropriate, before making any such decision. To the maximum extent permitted by law, the author and publisher disclaim all responsibility and liability to any person, arising directly or indirectly from any person taking or not taking action based on the information in this publication.

Contents

Foreword	Dr Niamh Shaw	*i*
Preface	Josh Richards	*vi*
Chapter 1	Questions	*1*
Chapter 2	The 101 Things	*15*
Chapter 3	A Real Bucket List	*45*
Chapter 4	Fear	*67*
Chapter 5	Attachments	*93*
Chapter 6	Sexy Space	*123*
Chapter 7	Getting Over It	*143*
Chapter 8	Little Martians	*165*

Appendix

Do. Or Do Not. There Is No Try	179
Sarah's Letter	185
Drawing - Welcome To Mars	191
Daily Habits	192
Writing	195
Mars One - Astronaut Selection Process	196
Mars One - Five Key Characteristics	198
Josh's Notes - Five Characteristics	200
Josh's Notes - Mars One's Book	203
Image Credits	205
Acknowledgements	206

Foreword

Dr. Niamh Shaw

I first spoke with Josh on St Patrick's Day in 2014. I thought that it would be a thirty-minute chat but it ended up being a conversation over three hours long. Skype was still a relatively novel way of communicating, I remember thinking how exotic it was that I was speaking with a random stranger in Australia.

About Space! On St Patrick's Day! As the rest of Ireland was downing green pints of Guinness, wearing shamrock, waving their shillelaghs and watching parades! As I was in my front room talking to someone in Australia about Space! And he was a ginger too!

This was kismet, had to be. He was the first person I spoke to, outside of my own small network in Ireland, and who instantly understood my passion for communicating space and my desire to get to space as an artist and communicator. It was the first time it felt as if I could genuinely find my own place in space, that what I cared about most in the world had relevancy. I had denied my passion for space for the first half of my life and these were my first tentative steps into realising this life long quest, to be a part of space, to finally feel as if my life had purpose.

I hadn't the first clue how to get to Space. It had all kicked off three years earlier, in 2011, making my first theatre show combining science with art. I was exploring particle physics and finding parallels between life choices and dimensions. And looking at my life choices, I realised that my childhood dream to be a part of space had not gone away - that it was still very much a part of me.

And now being an artist and no longer a full-time scientist, I had to pursue that. I needed to understand why space was so important to me.

The first two years on that journey of exploration were terrifying, I had no idea where to begin. I started following some NASA twitter accounts and wasn't sure what else to do. But apart from a trip to the International Space University in 2012, I hadn't met anyone outside of Ireland who was involved in space.

I eventually approached Blackrock Castle Observatory in Cork and told them that I wanted to explore the notion of being an artist trying to get to space. They loved it and made me their artist in residence. We applied for funding to make a theatre show and so in early 2014, I began making a show about my dream to go to space, called 'To Space'.

But still, I didn't know how to begin. For weeks I had been trawling through the internet, reading books, following Twitter accounts.
Late one night, in a desperate and lazy attempt to connect with people, I sent off about 40 LinkedIn 'space-related' requests.

No-one replied. Except for Josh.

The next morning there was a message in my inbox from him. And later that week we were on Skype, chatting on St Patrick's Day. The conversation was over three hours long because every question I asked, he generously and patiently answered, providing me with details that I had been so desperate to hear. He showed me around his room with his camera, sharing the posters and pictures on his wall. So much to see. I'd never seen a room like this before. A whole new world, a new perspective on how to live.

It felt as if I had stumbled upon an oracle on all things space, plus he was so friendly and keen to help. And when I checked out his website, I saw that our careers were so similar. He too was a communicator who shared a passion to communicate science to the public, we had both performed in Edinburgh, dabbled in comedy and we both loved Space.

I'm not sure if we Skyped again before we met the following September where Josh attended the opening night of my theatre show 'To Space', at the Dublin Fringe Festival. Two gingers who loved Space finally united! He was guest of honour at the opening night celebrations, everyone knew about him, he was the guy who had come all the way from Australia. He was the Josh that was mentioned in the play, he was the guy who introduced me to a ton of great people in the space sector. He was the guy that was heading to Mars.

He stayed with us while in Dublin. With my partner & I. Just him, a small green rucksack and a little tinny-sounding ukulele. The rucksack was faded and covered in patches from space missions, or countries he had visited. He had a travelling companion with him, she stayed in the spare room and he was happy to sleep on a therapy bed (I was learning a form of physiotherapy at the time). He seemed to need so little, and it had been a long time since I'd met someone like that. Had I ever met someone like that before? He reminded me that there was a time when I needed very little to be happy.

Cracks had begun to appear in my life by then, small at first, but since I'd begun on this path of Space, they were becoming harder to ignore and meeting someone with such simple needs seemed to expose the cracks all the more.

From then on we kept in constant contact. Catching up on Skype every couple of months, I called him the day he became one of the final 100 candidates for Mars One. I was coming off a flight from Amsterdam. My alighting passengers tut-tutted me for my loud squeals of excitement as I spoke to someone who had a one in 25 chance of getting to Mars!

Returning home from my first trip to the European Space Agency, life had already begun to rapidly change, those cracks back home soon becoming crevasses as I took another step closer into the space community. More and more difficult to ignore.

I secured a place on the International Space University's Space Studies Programme in 2015 - a nine-week intensive graduate programme which that year was held in Ohio. Another big step forward into the world of space. Navigating through that programme was difficult at times. I was a lot older than the participants, and as an artist, my mindset was often at odds with rooms full of technical engineers and scientists.

But a call from Josh would do the trick. As an alumnus of the programme, he would listen, advise, and encourage me to walk further outside my comfort zone. That it was okay not to be okay. That being outside your comfort zone was where I needed to be. I listened and really began to enjoy not having an answer. Not knowing. Not needing to know. It was a terrific experience that changed me forever.

I brought the theatre show 'To Space' to Adelaide Fringe in 2016, Josh brought his show 'Cosmic Nomad' too. He arranged for me to participate in a panel event at the World Science Festival in Brisbane. There I saw his show 'Cosmic Nomad', met his friends, and saw the green rucksack and ukulele again.

Reminding me again how simple life can be if we are brave enough and bold enough to keep walking forward. Walking our own path without compromise no matter what the cost.

Six months later I had begun my quest in earnest - everything from then on was about getting to space. Everything comes at a cost, but as long as we know that cost anything is possible. Then we can be the person we knew we were always destined to be.

I'm not sure if Earth will ever be enough for Josh. Mars may not even be enough. No matter where he ends up, he will always be kicking at the dirt questioning, challenging, exploring. An insatiable appetite for adventure with such resilience, but always with a deep understanding of humanity in all his adventures. A true maverick.

Josh is and always will be my friend, who I happened to meet through space. And that's what I'm most proud about. That and his green rucksack and ukulele which ultimately set me free.

Always and ever, a Cosmic Nomad!

Preface

Hi! My name is Josh Richards, and I'm an astronaut candidate for a one-way mission to Mars.

For the last eight years, I've been part of Mars One - an international not-for-profit organisation aiming to establish a permanent human presence on Mars. Founded in 2011 by two Dutch guys who are both smart and crazy enough to make it all happen, I discovered Mars One in September 2012 while writing a standup comedy show about sending people one-way to Mars for the Edinburgh Fringe.

As I sat in a little cafe by the English seaside, reading how four idiots would be shoved in a capsule and launched into the darkness of space at a tiny red dot 56 million kilometres away... I knew I needed to volunteer to be one of those idiots.

But this is not a book about going to Mars.
This is a book about living on Earth until then.

Signing up to be a Mars One candidate was an easy decision, but it's turned out to be the most disruptive event in my already chaotic life. While others have asked a relentless stream of questions about being shortlisted for a one-way mission to Mars, it's nothing compared to the torrent of existential questions I've had to ask myself: What will I do before I leave? Who will I have to leave behind? What if it all goes wrong? How can I have relationships on Earth knowing I'll eventually leave them for Mars? How do I want to be remembered after I'm gone?

Trying to answer those questions has catalysed more changes in my life than I can count, but it's the changes in my perspective that have been the most profound. Answering those existential questions about how I want to live on Earth before I leave for Mars has helped me see more of who I am, and decide who I want to be as a potential ambassador for humanity on another planet.

I don't know if Mars One will succeed in putting me or anyone else on the red planet - the challenges ahead are staggering, and there's no way to predict what fortunes the years ahead will bring the organisation or myself. But I do know that my life has gotten a lot weirder, the colours of everyday life have seemed brighter, and the opportunity to experience Earth fully has never felt more urgent than it has since I signed up to leave it.

So as you read this, don't just treat it as the entertaining ranting of some ginger space leprechaun trying to escape civilisation. Ask yourself the same questions I have, and figure out what you'd do if you knew your time on Earth is limited.

Because our time on this planet *is* limited, and it always has been.

Chapter 1 - Questions

Signing up for a one-way mission to Mars seems to encourage people to ask the most amazingly intrusive questions about my personal life. I guess meeting someone who's volunteered to leave the planet forces folks to reflect and ask tough questions about their own lives. So it's probably for the best that I'm pretty much an open book, and I'm happy to answer any awkward query anyone wants to throw my way. Still, it's endlessly amusing how often "Josh is going to Mars" fucks any other conversation. What's not so funny though is how quickly the conversation then devolves into pretty much the same questions, often coming from the same demographics.

For instance, middle-aged women without fail ask what my Mum thinks, and my less than polite answer is usually "Why don't you ask *her*?". I don't speak on my Mum's behalf, and she didn't sign up for a one-way mission to Mars - I did. There's also some sneaky sexism going on too since no one ever asks what my *Dad* thinks about it all. However, if I'm feeling generous, I'll usually say that both my parents are incredibly supportive *now*. In retrospect though, I really could have told them in a slightly gentler way than I did.

While I'd been living in the UK, I'd usually call my folks in Australia each Sunday afternoon to prove I was still alive. Since I discovered Mars One on a Thursday morning, I figured I had a few days to decide how to tell my parents I was moving to another planet.

Writing and journaling have long been the most effective way to turn my barely-coherent thoughts into something vaguely sensible, so I decided to try typing my way through it with a blog post. The writing just poured out of me, and the blog post was ready to publish less than an hour after I started it. Before I shared this ludicrous plan with the world though I still needed to speak to my folks, so I scheduled the post to publish the following Monday *after* our weekly call. Mum and Dad would get sufficient warning on my next ridiculous life-plan, and everyone else would hear about it roughly 18 hours later.

Except, we missed the Sunday phone call.

To this day, neither of my parents or I are precisely sure how it happened. It was hardly the first time we'd missed a Sunday phone call, but it's undoubtedly the most memorable. Sure enough, the blog post automatically published the next morning, and my phone rang around Tuesday lunchtime. When I first answered all I could hear was Mum sobbing something about Mars. Then suddenly Dad came on the line, and said he understood how excited I might have been... but it would have been nice if they'd had a little warning. You know, *before* telling the entire world I was signing up for a one-way mission to Mars.

These days my folks are my biggest supporters, primarily because I've dragged them along every step of the way. After the initial shock wore off, I quickly realised I'd need to return to Australia - the UK media would want to talk to *British* candidates, not antipodean imports. As terrifying as it might seem, I also knew I'd make a more significant impact speaking to Aussie kids as a supposed "role model". Cue *dry heaving*.

Moving back in with my parents at 27 might have been suboptimal, but at least it allowed them to see how vital this Mars-business was to me. Those first two years were a complete blur as I visited schools, did countless interviews, and performed shows for National Science Week while trying to write a book draft I'd eventually publish as *Becoming Martian*. Through all of this, my parents saw every day that I'd finally found something that made me leap out of bed in the morning. Something that gave me the life-affirming purpose they'd long hoped I'd find.

I know they wish I'd found something meaningful that was a little closer to Earth, but they also saw first-hand how lost I'd been in the decade after I left high school. Academia had been a frustrating and fruitless struggle, the senseless waste of the mining industry nearly drove me to suicide, and the military had me questioning my moral compass as I trained to pick up bombs and shoot people. Not only did Mars give me a sense of purpose unlike anything else, but ironically it'll keep me on Earth longer than most of my previous professions might have.

When I joined the Royal Marine Commandos at 23, I'd already accepted I probably wasn't going to make it to 30. A million different things might happen in the next 10-15 years that prevent me from going to Mars, but that's still 10-15 years on *Earth* preparing to leave. If I'd stayed with the Commandos in 2010, I'd have been doing regular 6-month deployments to Afghanistan where a million different things would have been conspiring every day to end me altogether. Even if I *had* continued to serve to the grand old age of 30, there's no promise I'd have done it with all my limbs, and a guarantee I'd have been even more deeply scarred on an emotional level than I already had been.

3 | *Questions*

So yes - Mum took a while to come around to the idea of me moving to another planet. But she also knows it's *my* decision, and she's a much bigger fan of me speaking in schools about leaving for Mars in the service of our entire species than she ever was about me potentially patrolling around some god-forsaken poppy field while trying not to step on an IED.

Awkward young men are the other group who regularly ask questions, and they're always shyly asking if there will be another call for astronaut candidates. The answer is yes: the first crew of four are just the start of a far more extensive Mars migration program. There's also no way to know if *any* of the candidates from the initial 2013 application period will be on that first mission at all. Every year of training is sure to bring new challenges and changing personal situations, and there'll be plenty of potential Martians that drop out over those 10-15 years of preparation. Mars One is set on maintaining an astronaut corp of 12 to 24 people, so applications will re-open every few years for those eager to take the place of any candidates who decide to stay on Earth after all.

The other "question" I regularly get from young men though is "I guess you don't have a girlfriend", which isn't so much a question as it is an assumption by sweaty-palmed dickheads. My usual response is that I've got *several* girlfriends, but we'll get to that later in the book.

Why?

Regardless if they're middle-aged women, awkward young men, or zany breakfast radio hosts; the one question I'm guaranteed to be asked by pretty much everyone is "Why?".

"Why would you give up your life on Earth?" Well, Earth is full of assholes, and I'd like to be a lot further from them.

"Why would you sign up to take such ridiculous risks?" You must have missed all the stupid shit I've done up til now for *no discernible reason*.

"Why would you *want* to live underground like a mole-person on some pitiless and desolate planet more than 200 million kilometres away, where the Sun isn't even half as bright as it is on Earth?" Who knows - maybe I'm just a risk-taking misanthrope with the lily-white skin of a ginger Martian vampire.

Why do people climb mountains, swim oceans, cross deserts, and sign up to leave the planet behind? To paraphrase George Mallory if you can't understand that something in us responds to the challenge and goes out to meet it, then you won't see why we go. Yes, Mallory died attempting to be the first to summit Everest, but he also died ATTEMPTING TO SUMMIT EVEREST. Stop trying to figure out if someone's "why" is "good enough" for you, and start celebrating those who aspire to do remarkable things.

Interestingly though it's only adults that ever ask "Why?". Kids get that doing something awesome is reason enough, so all they want to know about is how I'll shit in space. And they're sure to ask me about it every single time I speak to them about Mars.

Literally. Every. Time. They're usually in a school though, so they don't ask "How do you shit in space?" - it's always "How do you go to the toilet in space?" or "How does the toilet work?".

Eight years of school visits have taught me they don't care about where the urine goes though - they've already figured there's hoses for that. They want to know what happens to the brown stuff. I'll usually go off on a 10-minute explanation about how the toilet is the most complicated piece of equipment on a spaceship with specially designed vacuums, air jets, and a blender. That space toilets break down *a lot*, and it's the pilot's responsibility to repair them. And finally, that you have to be incredibly accurate when you're using it, or a turd can go rogue and escape the bowl. They might not say it in those words, but the intent is clear - if I visit a school, I'm *guaranteed* to have some kid ask me how to shit in space.

The other question I get from kids with startling regularity is "If someone dies on Mars, will you eat them?". I've lost track of how many times kids have asked me if cannibalism is on the menu, but strangely it's only six and seven-year-olds who seem keen to nibble on a crewmate. From what I can tell, kids younger than six haven't figured out that death is a thing yet. And while kids older than eight know that people die, they've also realised eating the body shouldn't be the *first* option for disposal. But it seems six and seven is the perfect age to be thinking about Martian cannibalism. Because *everything* in a spaceship gets recycled. Right?

Kids are always a little disappointed when I tell them we won't be turning Steve into a stir-fry. Well, at least not straight away. First, we'll have to liquefy his body through chemical cremation, then use him as fertiliser for the plants in the greenhouse, which we'll eventually eat once they're fully grown.

So you see kids we *will* be eating Steve, but he won't be the meat in our stir-fry - he'll be the bok choy. And *that* is the circle of life.

But while kids want to know how I'll shit in space or if the first crew will re-enact the Donner Party, they never ask why I'd want to go: they just intuitively understand how awesome it is. Or if going to space isn't their thing, they immediately tell me that they'd rather stay on Earth with their dog, cat, turtle, pony, trees, oxygen, or whatever other shit they (wrongly) believe is more important than going to *spaaaaaaaaace*! Teachers want to delve into "why", but their students don't give a shit about my "motivation" and don't judge my decision to sign up - they just know if they want to do it themselves.

That awful "Why?" question came up so many times when I first applied to Mars One, that I eventually had to come up with some sort of "reason". Because it seems the statement "I want to live on another planet" needs to be justified to adults.

So here it is: I want to live on another planet *because it'll be fucking awesome!* People whine that I'll "DIE on Mars", but they're missing the fact that I'll die **ON MARS!** You're all going to die on Earth just like every other tedious prick in history, but I have the chance to die on a different planet to the one I was born on! Hell, I might not even make it - I might die on the launch pad! Maybe I'll die in some sort of accident on the way there! Perhaps I'll die on impact and keep the red planet red! Regardless of how I die, at least I'd be doing it as part of a MISSION TO MARS!

I doubt I've convinced *many* people with this argument, but at least it stops most folks from continuing to ask that persistent "Why?".

Meeting someone who's volunteered for a one-way mission to Mars forces people to ask big questions about life, the universe, and everything. And their reactions to me seem to follow a rule of thirds. One-third of people think it's extraordinary - that what I'm involved with is the next giant leap for humanity, and that we're paving the way for a future where humans dedicate themselves entirely to exploring the universe we all share. These folks provide incredible support and are always eager to help spread the word on what we're trying to achieve by putting humans on another planet. They see that by developing technologies to survive on Mars, we can improve life here on Earth too. They know, just as Tsiolkovsky did, that Earth may be the cradle of humanity, but that doesn't mean we should remain here.

Another third just don't give a shit about any of it. And I can respect that - it's simply not their thing. I'm not here to preach the gospel of Ares: I'm doing something I care about, but you're free to do and care about whatever you like. In 1969 a journalist asked Pablo Picasso what he thought about the Apollo 11 lunar landing, and was surprised when Picasso replied: "It means nothing to me". Nor should it have meant anything to him - he was busy drawing boxes and calling them people, while actual people were building actual boxes and flying them to the fucking Moon. What we *should* ask is why anyone gives a shit about what Picasso thought about the lunar landing? He had his thing, that thing was painting and *not* exploring the Moon, and that is entirely okay.

I don't need random strangers *telling* me they don't give a shit about missions to Mars though. A few years ago I was stage managing a theatre play, and several weeks into the run the director suddenly walked up to me just before the show and said: "I heard you're going to Mars."

I smiled and nodded, and he responded "That's cool... I guess..." then walked away without ever mentioning it again. It's not like I asked what he thought, but thanks for sharing anyway, dipshit. Years later, I'm still grateful that my aspiration to become one of the first people to walk on another planet once received tepid support from an amateur theatre director.

And finally, there's the third of people who *fucking hate everything about you and the idea of ever sending people to Mars*. They'll keep screaming the radiation will kill you in a matter of months, long after you've shown them the data that proves otherwise. They'll keep telling you humans will never adjust to the isolation and cramped quarters, while politely ignoring countless crews who overwinter in Antarctica or live on nuclear submarines that only surface to take on food. They'll tell you humans will never adjust to the slightly longer day on Mars, forgetting about the teams of speleologists who spend months underground exploring new cave systems. Or they'll tell you we can't go to Mars until there are new propulsion technologies to get us there in less than six months, even though people have been spending stretches of *seven* months or more in space since 1984.

The folks who get angry about me trying to go to Mars one way are **great** though because you know they're passionate about exploring space - they just don't like how **I** am planning to do it. If they didn't care about space exploration, they wouldn't be upset that I'm trying to do it the "wrong" way. The folks who get mad about human spaceflight are pretty hilarious too though: How dare we waste money on space exploration when people are homeless and starving?! We could build hundreds of hospitals for the money you'd spend going to Mars! It's always fun to ask people who see space exploration as a waste why we have to choose between caring for people on Earth and sending them to space.

9 | *Questions*

Especially when there never seems to be a shortage of money to build weapons that kill people.

Buuuuuuut if I'm honest, I don't care about convincing those people either. I could spend all my time and energy justifying why going to Mars is worth it, and pour everything into arguing that humanity needs to live beyond Earth to survive the next Extinction Level Event. Or I can just fucking *do it*. The folks who get upset about me trying to live on another planet don't hate *me*, even when the attacks can feel intensely personal. They're just upset by something that challenges their world view, and the candidates who've signed up for a one-way mission to Mars are the representatives of that threatening concept.

I've found if you don't take it personally - choosing to be kind to the sceptics by answering their questions without assuming any malice - they'll usually realise you've asked yourself the same questions long before. There's often a dawning realisation that they've developed an opinion of you and what you're doing without having all the facts themselves, and it's a revelation when you point them to research they think you haven't done. You probably won't win them over, but at least they'll usually stop spouting half-assed arguments like it's some sort of academic pissing contest that you never asked for.

What you should definitely *not* do however is <u>laugh</u> at the folks who *do* want you to take their attacks personally. If someone goes to the effort of setting up a dedicated hate-page on Facebook named "Josh Richards - Space Fraud" so they can rage impotently at you in the comment section, then it's "very important" to be upset and take it all *deeply* personally.

Do not under any circumstance start following their hate-page, message them to say how good a job they've done editing your photos with MS Paint, and then publicly encourage your fans to like their hate-page too. I've since been informed this is *not* the correct response to these situations, and I may in fact have helped create a future serial killer.

I still love you anyway, random internet troll

Sarah's Letter

It's also quite okay that a third of people get bent out of shape by something as utterly ridiculous as a one-way mission to Mars, because it often forces them to question their core-beliefs and decision making. Hell, Mars One forces people to question *everything*, especially when they're face-to-face with some idiot saying "I wouldn't want to come back, even if I could". Suddenly you have to ask if all the shit you *thought* was important - your house, your job, your partner, your pets, your collection of voodoo dolls made from the hair of your enemies - actually means anything in the grand scheme of the universe.

If you had one chance, one opportunity, to seize everything you ever wanted, and burn it as rocket fuel on a journey to Mars... would you take it?

Although, it turns out I don't actually have a choice in any of this. I *have* to do absolutely everything I can to get to Mars. Not because of some misguided compulsion to explore the universe, a warped obligation to act in the highest and best interests of humanity, or because I've deluded myself into believing it's somehow my "destiny" to be one of the first people to live on another planet. No, I have to go to Mars because Sarah Young has promised to hunt me down and brutally murder me if I don't.

I've known Sarah for years through the Australian standup scene, and she's always had a delightfully warped take on what constitutes comedy. I say delightful because it's the kind of comedy where someone says something truly sickeningly awful, everyone laughs because "Oh haha, there's no way she's serious about this - that'd be psychotic!". And Sarah always takes a *little* too long to follow up with "Ha, yup. That's right. A joke." while maintaining a perfect deadpan. I'm still not entirely sure if her one-woman musical sock puppet show about serial killers was a comedy or a DIY workshop.

Sarah was halfway through her run of terrifying audiences with nightly renditions of "Dahmer and Greg" at the 2015 Perth Fringe festival when Mars One selected 100 candidates for the round 3 shortlist, and she was beside herself with excitement when she learned I'd made the cut. A few days after I shared the news, she unexpectedly pulled me aside after her show and solemnly provided me with what she referred to as her "Letter of Recommendation" to Mars One. The five pages were folded together in thirds, with "FOR JOSH IN TIMES OF DOUBT" written in bright red sharpie on top.

Riding on a wave of excitement, I promised Sarah I would *never* need it: I'd dedicated everything to this, so there was no way I'd back out! She politely nodded and made me promise to open it if I ever did start to find doubt creeping in.

Sure enough, just a few days later, it all started to hit me. Here I was shortlisted from over 200,000 people who'd first started the application, and what did I - a ginger idiot who tried to make people laugh for a living - have to offer a mission to Mars? The 100 candidates included nuclear engineers, spaceship designers, astrophysicists, biochemists... and what was I? Some shoddy physicist who barely scraped through with a Bachelor's degree? Some angry ginger kid who joined the military as a non-commissioned grunt to blow things up? Someone who bumped around a few different jobs without truly committing to any of them, got into standup as onstage therapy rather than to develop a craft, then signed up for a one-way mission to Mars because I thought it'd make a decent comedy show. There was no way I belonged in this incredible group of astronaut candidates and future Martians!

Sarah's letter called to me, so I read it as I'd promised I would in times of doubt. Within a few paragraphs, I was smiling as she spelled-out just how much I had to offer Mars One. For all my self-doubt, she reminded me that I *was* profoundly qualified for this. That I had a talent for thinking strategically, could handle stressful situations to solve problems, and that I had been tireless in my pursuit of something that would have an extraordinary impact on future generations. Sarah told me that she'd never said goodbye to a friend who "left the fucking planet before", how writing that sentence had made her laugh, and that she didn't know many people willing to sacrifice so much "just to advance the human race that little bit further".

If she'd left it there, I probably would have kept Sarah's letter for a few years then forgotten about it. But Sarah didn't stop there. After two and a half pages of expounding on what an incredible thing I was doing and how much I had to offer humanity by volunteering for a one-way mission to Mars, her tone suddenly changed:

"And if for any reason at all, you chicken out of this opportunity, here is the process of what I will do. I will dedicate a large portion of my life to hunting you. You will not even know I am there. You know the way a Great White Shark stalks Josh? It sinks its dorsal fin under the water so it won't be detected. If for some reason you pass this opportunity in the face of self-doubt or any other equally lame reason, here is how you will be murdered by me:

Transcript from the Dateline Mystery Episode 99 "Mars or Death" that was made in the wake of your murder.

Voice Over:
...But Richards had no earthly idea what he was about to face."

Which goes on for another two pages as Sarah describes strangling me with a garrotte in a bathroom at a comedy club, with my body found by an open mic comedian. At the same time, she plans to escape from the police with my decapitated head - hiding it in the luggage of a Mars One crew member so at least part of me can fulfil my "destiny" of getting to Mars. If I'm perfectly honest, being found by an open mic comic is probably the most upsetting part of the whole scenario, although having my death rattle described as sounding like a "shrieking weasel" is a close second.

Two things are certain though: I'm wholly committed to becoming one of the first people on Mars, and I'm never setting foot in a comedy club bathroom ever again.

Chapter 2 - The 101 Things

So if you've committed to dying on a cold and desolate planet instead of being decapitated in a cubicle at Lazy Susan's Comedy Den, what exactly do you do with your remaining time on Earth? Do you try to make your limited days on this planet *mean* something? Plenty of people talk about having a "bucket list" of things to do before they die, but most don't have an actual *deadline* for it. It's just some whimsical list of dreams they write up in their 30's, revisit during a mid-life crisis, then grow slowly more bitter about as the years slip away into the twilight of death's warm embrace. Even I have to admit that until Mars One and Sarah's death threats came along, I was probably heading down the same path.

Rather than just writing up a dreamy bucket list generously padded out with my wildest fantasies, I've had something tangible to guide me: Richard Horne's "101 Things To Do Before You Die". My best friend from high school had given me a copy just before we left on a round-the-world trip through the UK and North America in 2005, and I'd immediately attacked it with a healthy mixture of unjustified zeal and incompetence.

Right from the get-go, there were roughly 20 of the 101 entries I could tick off the list as completed: I'd been scuba diving since I was 12, I'd milked cows growing up, had seen both a solar and a lunar eclipse, and had learned to play a musical instrument. Although I'm not sure anyone can honestly call my ukulele rendition of *Werewolves of London* particularly "musical".

The US trip offered plenty of opportunities to tick off even more entries. Paul and I bungee jumped in Canada. We threw a dart into a map of San Francisco, then visited the Starbucks at the corner it landed on. And we sat terrified in the back of a sports car as it hit top speed, flying through the streets of Baton Rouge as the Barney Rubble look-alike driving told us "Doncha worry boys! Ahl git yawl thayre on taym!".

After the trip though, my 101 things book quickly fell into neglect. In the years that followed I checked it sporadically, even occasionally ticking things off if the opportunity was there. But the urgent awareness of my mortality the book had encouraged soon faded: what was the point of trying to tick things off someone *else's* list anyway? I didn't expect to make it to 30, but that didn't mean I was going to try and see the world's tallest buildings before kicking the bucket, just because some book told me to.

Discovering Mars One changed all of that in a heartbeat. Instead of arbitrarily contemplating a list of things I might like to do before I died, Mars One suddenly gave me a clear deadline. I might not be dying, and my departure date is still at least ten years away, but it became instantly clear that one way or another my time on Earth is - and always has been - incredibly limited.

So if you've been abruptly gripped by a sense of your own mortality, and decide to tackle an arbitrary list of personal challenges as a distraction from a crushing existential dread... then it's crucial to have a plan to tackle that list! For instance, some of the "101 Things To Do Before You Die" are simply impossible to complete once you leave Earth: last I checked *none* of the world's biggest rollercoasters are on Mars. Conversely, some entries will be even *easier* to do living on the fourth rock from the Sun.

For instance, reading *One Hundred Years of Solitude* by Gabriel Garcia Marquez from the list of all-time greatest books will be a lot easier when you're living on a cold-lifeless rock 56 million kilometres away from the rest of humanity.

To date I've completed 73 of the 101 entries, so here's how I'm tackling the remaining 28.

5 – Make a Discovery [Mars]
To be fair, this is going to be a whole lot easier on an otherwise uninhabited planet. On Earth, there are famous explorers like Sir Ranulph Fiennes to compete with over discoveries. But on Mars everywhere you go, you'll be the first. The first to discover a water aquifer on Mars. The first to learn how our bodies adapt to 38% gravity. Maybe even the first to find Martian fossils or microbial life. With an entire planet to explore, there is no shortage of discoveries waiting to have some idiot stumble onto them. I just hope I don't complete this entry by climbing down a lava tube and "discovering" a facehugger as it latches onto my helmet and melts its way through the visor.

13 – Meet Someone with Your Own Name [Earth]
With 7 billion people on Earth, it'll be a whole lot easier to meet someone else called "Josh Richards" on *this* planet before I leave for the red one. That said, I don't want to meet up with any old "Josh Richards" - I want to meet an equally awesome one. At the 2012 Edinburgh Fringe, someone confused me with a Welsh actor with the same name, but it turns out he was born Mark and *changed* his name to Josh... so he doesn't count.

There's some 18-year-old Canadian musician who should probably spend less time on TikTok and more time getting a damn haircut, as well as a NASCAR driver who seems like a pretty decent guy...

Except that NASCAR guy is the reason I can't register "www.joshrichards.com". Which is fine. **TOTALLY** fine. I've got "www.joshrichards.space" instead, which is more appropriate for me anyway... Honestly, it's **FINE JOSH - YOU GO AND TAKE THE BEST WEB ADDRESS TO PROMOTE YOUR BIG GO-FAST CAR. SHUT UP. WHATEVER.**

However, because I'm always on the lookout for the latest news about my own stupidity, I've set up a Google alert that emails me whenever "Josh Richards" and "Mars" appear together in an online article. In mid-September 2017 I suddenly started getting emails about a "Josh Richards" who survived a crash-landing during a wingsuit BASE-jump in northern Italy *without ever opening his parachute*. The air conditions changed so quickly during the jump he didn't even have time to pull the chute before he slammed into a bunch of trees at 150km/hr, but then *walked away* from the crash site with a broken leg and a few fractured ribs. Now *that* dude sounds like a genuinely awesome "Josh Richards".

Oh, and it also turns out he's a mining mechanic that lives about 20 minutes away from my parents' house in Perth. So I strongly suspect I'll be ticking off this 101 entry reasonably soon, and I **won't** be doing it by meeting a name-changing actor, a website-stealing NASCAR driver, or some scruffy-looking, nerf-herding TikToker from Canada.

14 – Ride the World's Biggest Rollercoasters (5 of 6 complete)
[Earth, obviously]

I've never really understood the appeal of rollercoasters. Being strapped into a cart that's then thrown down a track as you're tossed around inside has never struck me as particularly "fun".

In 2016 I was halfway through my last global standup tour, and between Cleveland and Cincinnati I figured I might as well call into Cedar Point, aka "Rollercoaster Capital of The World", to ride the fastest of the six rollercoasters listed: Top Thrill Dragster.

There are only two words to describe the experience: Holy. Shit.

This thing accelerates you from zero to nearly 200 km/hr in less than 4 seconds. You pull about 1.4g horizontally then twist 90 degrees vertically. The cars run out of steam 120 metres straight up where you pause briefly, have a look around Ohio for a bit, then suddenly race back down again and stop back where you started. The entire ordeal is over in less than 15 seconds, but it's enough time to discover your whole face can (and will) flap in the wind. Now I still think rollercoasters are stupid, but after Top Thrill Dragster I'll tentatively accept why other people might think they're exciting.

That trip to Cincinnati also took me to Kings Island, where I'd hoped to ride "Son of Beast" from the official list, but it'd closed in 2012 after hospitalising 28 people. Instead, I rode the original "Beast" at the same park. It might be the longest wooden rollercoaster in the world, but if I'm honest all the "Beast" did was make me question why *anyone* would ever *choose* to build a noisy, rickety rollercoaster out of wood instead of something sturdy like, I don't know, *steel*?

During another visit to LA, I spent the day at Six Flags Magic Mountain to tick "Superman the Escape" off the list as well. It was alright, I guess. Although in retrospect, I might have remembered the experience a little more fondly if the 10-year-old kid sitting behind me on the ride hadn't hurled bright blue Slushie mix halfway through the zero-g section. But that's not the ride's fault - that's just bad parenting.

However, I've probably completed the three rollercoasters that were the most out of my way. The remaining three are in Japan, London, and Florida where I have friends to visit. So now it's merely a case of calling into amusement parks next time I'm in the neighbourhood, and then finally crossing this entry off the list!

19 – Master Poker and Win Big at the Casino [Earth]
Card games are an excellent way for a Mars crew to bond and pass the time. However, we're also supposed to *trust* each other, and playing poker is the perfect way to fuck that up. Personally, I think casinos are the biggest pits of despair on *this* planet, so I'm not rushing to set one up on Mars.

Thankfully though, there's already a barren wasteland with plenty of gambling to visit here on Earth, which goes by the name of Las Vegas. A city that runs on gambling in a desert sounds precisely like my vision of Hell, but I *have* long dreamed of driving a campervan along Nevada's State Route 375 - aka the "Extraterrestrial Highway" - and parking in front of Area 51. I could overnight in Vegas to play poker, tick off "#88 Get Married Unusually" by having a drunk and overweight Elvis slur through my wedding to a half-naked jelly-wrestler. Next morning Elvis and I would drive out to Area 51 and wait to be abducted by aliens or the US Air Force. If I'm going to tick this entry off, I may as well get a trip to space out of it too.

23 – Get Arrested [Earth]

I'll be the first to admit that I've done plenty of dodgy shit through the years, but I've always taken particular pride in not getting caught. That'll have to change while I'm on Earth if I'm ever going to complete this entry though.

The first people living on Mars can't even be arrested for a noise complaint because the nearest Five-O will be more than 50 million kilometres away.

So to complete entry number #23, I'll need to work out how to be arrested on Earth - ideally for something hilarious, but also not so severe that others will hesitate climbing into a capsule to Mars with me. Well, at least no more than they might already. The obvious option is to try to get arrested for breaking some completely ridiculous law - something like visiting Portugal and peeing in the ocean, or holding a salmon "suspiciously" in England. In Alabama, it's understandably illegal to throw glitter, but you can also be jailed there for wearing a fake moustache that causes laughter in a church.

As tempting as it is, I'm not sure I ever want to *be* in Alabama, let alone get arrested there. But if I am, you can be sure they'll arrest me running through a church wearing a giant moustache and throwing glitter. The local sheriff probably won't find it amusing, but I'm certain Nietzsche would appreciate the cheerful nihilism of it all.

If I'd been smarter about this, I probably could have ticked this entry off years ago by having too many potatoes in my backyard! Up until 2016, you could still be charged under Western Australia's "Marketing of Potatoes Act" of 1946 for selling, delivering, purchasing or taking delivery of more than 50kg of potatoes unless you were a member of the Potato Marketing Corporation or a registered agent.

There were legitimate "potato inspectors" who would stop and search your vehicle if they suspected you were carrying too much of The Martian's favourite food. To this day, I still wish my Mum and Dad had run an illegal potato trading racket during my childhood.

Of all the "101 Things To Do Before You Die", this has turned out to be one of the most challenging to find a way to complete. Rather than actively setting out to be *deliberately* jailed though, I'm now hoping I'll be confused with someone else and get *accidentally* arrested.

Every time I've flown into the US since 2001, I've wound up in "secondary screening". Homeland Security have told me it's because my fingerprints are similar to someone on a watchlist. Or maybe *I* am the one on the watchlist, and they just don't want to tell me they've bugged my bags when they hand them back. Either way, I know I'm in with a chance.

24 – See a Rocket Launch [Earth]
As someone who wants to die on another planet, you'd think I'd have seen at least one rocket launch in person by now. I guess I could just wait till I *go* to Mars and say that I saw a rocket launch *from the inside*, but I suspect that's another experience altogether. It's not for lack of trying to see a launch though.

After three days of delays, I arrived in Florida TWO HOURS after Thaicom-8 launched out of Cape Canaveral. There was a little solace in getting to see the first-stage booster return to port a few days later though. I also spent a week in a campervan trying to see Iridium-6 launch out of Vandenberg, before hearing the launch had been pushed back to the day after I had to fly out of the US.

Of course, as soon as I arrived back in LA, news then filtered through that the delay had been someone forgetting to unplug a lead, so SpaceX was moving the launch *forward* again. For a few seconds, I seriously considered racing 300km back to see it after all. But one glance at the weather forecast told me the entire launch was going to be blanketed by dense fog anyway - maybe next time.

In fairness, if I'm going to see my first rocket launch, it should be something *truly* incredible. There was an enormous temptation to fly across the planet to see Elon fire his cherry red mid-life crisis into space on the first Falcon Heavy launch. Instead, I resisted and lamely watched it on Youtube like most of humanity.

If I haven't seen a launch by the time SpaceX's Big Falcon Rocket (BFR) is ready though, there is no way I'll be able to resist the urge to book flights and witness the inaugural launch of the biggest rocket in history.

Launch delays are pretty standard, unless of course you want to traipse out to the steppes of Kazakhstan in a blizzard to see a Soyuz take flight. Because those crazy Russians will launch in absolutely anything. A week-long trip hanging out with Kazakh goat-herders might sound tempting, but I'd prefer to wait until BFR's maiden launch out of Cape Canaveral is looking likely. That way when the inevitable delay comes, I can hire a campervan for a few weeks and learn to cave dive at Devil's Den, hang out with the Manatees in Crystal River, and try to remain cynical in the face of Disney's unrelenting optimism.

It's a small hell after all

Plus, if it looks like a 4,400-ton rocket is going to malfunction and level NASA's most famous launch facility in a blast bigger than the bomb the Americans dropped on Hiroshima, then at least I'll have an underpowered motorhome to try and outrun annihilation.

25 – Capture the Moment in an Award-winning Photograph [Mars]
Photography is not something I've ever had a real passion for. I grew up using high-end cameras and became reasonably good at animal and underwater photography. But whenever I've carried a camera, it's always felt like an obligation - something I *had* to use rather than just enjoying the sights for myself. After travelling out of a backpack for a few years, I realised the camera in my phone was doing all the photography I wanted. So I ditched the DSLR I'd been lugging around the planet and I've never looked back.

If I dug through some of my old shark diving shots, I suppose there'd be something that could win a photography award. But as usual, I'd much rather complete this entry by doing something awesome instead of just ticking a box. So rather than competing against every camera nerd on Earth, I'm just going to get the Mars crew together for the first selfie on the red planet, send it back to Earth, and then win *all* the awards. Stupid faces will be encouraged, but anyone caught dabbing in the background will be immediately pushed out an airlock.

27 – See an Erupting Volcano [Earth]
Like many bright and cheerful young kids, I grew up with an almost obsessive interest in space, dinosaurs, and the ancient Roman city of Pompeii being consumed by pyroclastic ash in 79 AD. Yup, **a totally natural thing to be obsessed with as a child.**

I'd still love to explore the ruins of Pompeii and maybe even climb Mount Vesuvius, but the chances of it exploding while I'm there are pretty slim. Sure, it's technically still active and will eventually cancel Christmas for the 600,000 people living in the red zone, but there hasn't been a major eruption since 1944. If I'm going all the way to Italy to gleefully run through the ashes of former people at Pompeii though, Stromboli is just a 4-hour boat ride away, and *is* continuously throwing out lava bombs.

I'd also love to do a night dive with the manta rays at Kailua Kona in Hawaii sometime, so it'd be easy to also do a lava boat tour while I'm there. And if for some reason I ever feel compelled to visit the Congo and risk altitude sickness to stare at the gaping mouth of Hell, then I can always put Mount Nyiragongo on my "To Do" list as well.

31 – Experience Weightlessness [Earth]

If I'm spending seven months hurtling through the darkness of space to get to Mars, I'm going to have plenty of time for zero-g ballet. Every great performance requires plenty of practice before the big event though, so my future Martians and I will be training to move around in space by flying up and down on "parabolic flights". You get in a hollowed-out passenger jet that climbs to about 32,000 feet, which then nose-dives 8,000 feet to provide everyone inside about 30 seconds of weightlessness. The plane then climbs back up to 32,000 feet to do it **all over again**. And after about 30 of these up-and-down cycles, you'll probably work out why it's lovingly referred to as the "Vomit Comet".

Zero-G corp have been running parabolic flights across the US for the public since 2004, and while shelling out USD$5000 for a ride *is* an option, what I'd really like to do is visit the cosmonaut training centre at Star City in Moscow so I can float around inside a *Russian* vomit comet! That way I'll get to see where Yuri Gagarin and every cosmonaut since has trained, bounce weightless inside a Russian cargo jet, then head out on a vodka tour through Moscow to wash the taste of despair from my mouth.

32 – See the Aurora Borealis [Earth or Mars]

Seeing the Northern or Southern Lights has been the entry I've been most excited about completing ever since I was given the 101 Things book in 2005. It's also the entry I've had the least luck with completing. Iceland had massive aurora the month *before* I arrived, but not even a flicker during the week I spent camping there - go figure. The clear skies over Tasmania exploded into light the weekend I started an artist residency on Flinders Island... but I didn't hear about it until the next morning, as the cloud cover settled in for the next three weeks.

At this point, I figure I'll have to either go on a dedicated aurora tour, or live somewhere that has aurora so regularly you need blackout blinds to sleep at night. Scuba diving in Antarctica has been a lifelong dream, and I'd love to torment Santa in Lapland before riding off into the midnight sun with his reindeer. But given that I'm an anti-social, sun-hating penguin, I may as well spend most of my remaining time on Earth living somewhere cold, clear and relatively civilisation-free anyway. The outskirts of Hobart look like a good Australian option, and a friend owns a pub in Nova Scotia if I find myself in Canada too. Either way, sitting outside in the snow watching the night sky while drinking with a bunch of rowdy Tasmanians or Canadians sounds like my jam.

I'm pretty sure I'll tick this entry off sooner rather than later, but if for some reason I don't see an aurora before I leave Earth there are still a few options. For starters, I don't necessarily have to see the Earth's aurora from *Earth*. We'll need to stay in Low-Earth orbit for at least 24 hours before making the big push on to Mars, and astronauts on the International Space Station regularly report flying *through* the Earth's aurora. For several years I was also under the impression there were no auroras on Mars because it doesn't have a global magnetic field. But it turns out it has faint but *planet-wide* auroras.

The vivid red and green sky shows we get on Earth are caused by the solar wind hitting oxygen and nitrogen in the atmosphere after being concentrated by this planet's magnetic poles. With a wispy-thin atmosphere made mostly of carbon dioxide and almost no magnetic field, the auroras on Mars will be incredibly faint and blue. So while I could technically complete this entry by climbing out of my underground hermit hole on Mars one night, a Martian aurora is barely worth the extra radiation exposure.

What *would* be worth the cancer risk is seeing aurora over the solar system's outer planets. Imagine sitting on the ice of Europa and watching an aurora over Jupiter? Or an aurora over the enormous hexagonal cloud pattern above Saturn's northern pole? If someone ever offers me a ride off Mars, then fuck ever coming back to Earth - there's way too much other cool shit to see in the rest of the solar system before I ever think about coming back to a planet named after dirt.

36 – Visit Every Country/Continent [Earth]
Whenever I meet someone who proudly proclaims they've "visited 35 countries", my eyes immediately roll back in my head and I start making this weird gurgling sound. I've been lucky enough to see a lot of this planet already, so I'm not in the slightest bit impressed by your "number". I want to hear about the time a camel-spider bit you in Senegal. Or how you had your passport stolen on the Trans-Siberian railway. Or the time you ate dodgy hash brownies in Amsterdam and thought you could see the future and the past at the same time. It might be easy for some douchebag who wants to die on Mars to say, but please don't travel around this planet just to collect flags. It's tedious and about as tasteful as adding up how many times you've had sex that year, then proudly announcing it at your family's Christmas dinner.

I do want to see all of Earth's continents though, and not just from space. While I flew over Antarctica in 2003 to see a solar eclipse, a lot of folks have told me it doesn't count unless you've walked on the ice. If we're pedantic though, I don't think you can say you've genuinely experienced Antarctica unless you've gotten hypothermia while watching a Leopard seal shred a penguin.

Most Antarctic tours leave from Ushuaia, near the most southerly tip of Argentina. So if I head off on my polar adventure to witness penguin-puree from there, then I'll also be able to tick off South America - the only other continent I haven't visited. If flying over Antarctica doesn't count though, then passing through a single Patagonian port hardly qualifies as "visiting" South America.

The Eastern half of the continent has never personally held much appeal, but walking the Inca trail to Machu Picchu has been an ambition since I read about it as a kid. Rather than just quickly racing along a tourist trail to see an abandoned mountain-top city as a pit-stop on my way to Antarctica though, I want to spend at least three months exploring the parts of South America that appeal most. A long hike through the Andes to climb Aconcagua would be an extraordinary experience, and I'd have to organise a boat to the Galápagos and chat to the iguanas too. And while it's *technically* Central America, there's no way I could ever fly within 1000km of Costa Rica without stopping to stare dumbly at the sloths and their ridiculously adorable faces.

48 – Be Present When Your Country Wins the World Cup [Earth]
Personally, nationality is an out-dated and meaningless way of defining who we are... which is also probably why I've never understood international sports competitions. It genuinely blows my mind that anyone would pay money to cram into plastic seats with thousands of others to watch sportsballers from one arbitrarily defined patch of Earth try to "win" an artificially-created game against sportsballers hailing from some other piece of dirt. "But it's all about the roar of an excited crowd and the electric atmosphere". Thanks, but it may shock you to learn I didn't volunteer for a one-way mission to a cold and desolate planet because I *like* crowds.

The only time I'm going to worry about "atmosphere" is when the rest of the Mars crew threaten to shove me out an airlock for being a fun-hating misanthrope.

Thankfully this particular entry was updated to include seeing your country win a gold medal at the Olympics, so I no longer have to work out how to endure two hours at a soccer or rugby final. If I'm ever going to suck it up and tolerate a crowd to see Australia win a gold medal in something, it'll probably have to be watching someone swim laps at the Olympics. COVID-19 may have pushed the games in Tokyo back a year, but 2021 might be my chance to finally tick off this entry while simultaneously teetering on the edge of emotional implosion whenever I'm hustled onto a train or into an aquatic centre. At least afterwards I'll be able to immediately jump on the Shinkansen bullet train to Sapporo and disappear into a forest somewhere on Hokkaido. With a bit of luck, someone will have secretly built that machine from *Contact* that Jodie Foster rides into a wormhole, that way I can fuck off to another galaxy while I'm at it.

52 – Read the Greatest Books Ever Written [Mars]
If you were planning to live underground on a frozen wasteland of a planet, would you spend your remaining decade on Earth reading things like *The Master and The Margarita*? No - you'd focus on entries that require wide-open spaces, that I *can't* complete breathing recycled farts in an aluminium tube on another planet. To date, I've read 36 of the 85 books on this list, so I'm certainly not opposed to reading the classics. But I also think this entry is a perfect example of "Shit to do on Mars Because You Can't Play Outside".

There's still every possibility I'll finish all 85 books to complete this entry before I land on Mars though, given we'll be drifting through the darkness of interplanetary space for seven months to get there.

Most of Mars One's training will also involve living for months in a simulation habitat, while the selectors watch us through CCTV to work out if any of us are going to get stabby. So if I do make it through the next selection round, it might pay to read *1984* <u>before</u> training begins and I have to live in a claustrophobic box under 24-hour surveillance.

60 – Take Part in a Police Line-Up [Earth]
I promised that I wouldn't judge the entries on this list... but how the hell am I supposed to do this? Police identity parades are practically non-existent in Australia and the UK, because it's so much cheaper and easier to ask a witness to identify someone from an interview video. **No one** wants to do line-ups: not me, not the cops, and certainly not the dude who looks like me that got arrested trying to smuggle a raccoon onto a passenger jet. Parts of the US still do them, so I guess I could volunteer for an identity parade there. Weirdly enough though, I'm also not super keen on being potentially confused with someone else in a country that houses over 20% of the world's prisoners.

So just like "getting arrested", I'm not putting any time or effort into completing this entry. The cops don't want or need any of us hanging around and eating their doughnuts or drinking their coffee. That said if there *is* some ridiculous opportunity to do a police line-up then absolutely sign me up. Fingers-crossed there'll be a traumatised flight attendant behind the one-way mirror, sobbing to the cops that it was "#2 - the Australian" they saw unleashing a trash panda into first class.

62 – Join the Mile-High Club [Earth]

Hiring a private jet and someone to fly it for a few hours isn't really in my budget, and small charter flights feel bumpy enough without two (or more) people repeatedly changing the aircraft's centre of mass. I *can* fly a plane so I guess I could always multi-task, but I also know the guy who invented the autopilot crashed during "flying lessons" with a female pilot-in-training: both were plucked from the ocean safely but also completely naked. So it seems my only immediate option for this entry is to try to get cheeky on a passenger jet.

Trying to join the mile-high club on a crowded flight is much like attempting a mission to Mars. You have to plan ahead while still being open to spontaneous opportunities, commit fully to your plan without being overwhelmed by "Go" fever, and choose your "crew" wisely... Although that doesn't mean it has to be the **flight** crew. I'd also like to add that I'm keen to avoid completing #13 at the same time, although this *would* be a hilarious way to get arrested.

After extensive research and study, it looks like the best option is a long-haul flight that departs at night and flies West to stay in the darkness. Most people onboard will try to sleep through the extended "night", providing you and your accomplice (or *accomplices*) some extra cover for your airborne activities. There's a bundle of different flights that depart Los Angeles in the middle of the night and arrive in Sydney 15 hours later, so maybe I'll tick a box or two over the Pacific sometime. Crossing the International Date Line and losing an entire day in the process isn't ideal though, so it might be better to fly direct from Perth to London instead. That way you can woo them at dinner in Northbridge, spend 17 hours flying out of the world's most isolated capital city, then take them out for English breakfast in Soho the next morning.

That's not to say I couldn't complete this entry in *space* though, however joining the mile-high club in microgravity is sure to have a few extra layers of difficulty. BDSM fans will rejoice because the first thing you'll need is some sort of restraint system to stop yourselves from flying apart with each thrust. It's not just straps and safe words though - architecturally-minded perverts have been designing tubular habitats for decades, while others have spent their entire career "researching" specially designed "hugging" suits that zip onto each other.

Even if you *do* have all the necessary hardware and some willing research participants though, the guys are still likely to encounter "hydraulics issues" no matter how mentally excited they might be about getting down in orbit. Without gravity pulling blood towards the legs, it shifts to the upper body where it smooths out wrinkles and makes chests look bigger. But this simultaneously leaves less blood available to help transform the pants department beyond the initial "wet noodle" phase. I'm not saying it's impossible, just a lot more challenging than it'd be when gravity is helping the blood rush where the brain wants it to be. But in the wise words of Dr Ian Malcolm "Life, uh... finds a way".

With 38% of Earth's gravity, I'm hoping there won't be as many hydraulic challenges on Mars as there are in zero-g. Although I'm not sure it'll still count as "joining the mile-high club" if you're living underground in a hobbit hole on another planet. You might be more than 34 *million* miles above the Earth, breathing recycled air and living in something shaped like the cabin of a passenger jet. But banging a crewmate while you're living like mole-people under 5 meters of Martian dirt doesn't quite strike me as *flying*.

63 – Make the Front Page of a National Newspaper [Earth or Mars]
Most folks probably see making the front page of a newspaper or appearing on primetime TV as a **huge** deal, and something they'd want to share and discuss relentlessly. And yeah - being part of the global media coverage around a human mission to Mars *does* provide a sense of recognition for my small part in trying to help humanity become a multi-planetary species. I also genuinely believe it's vital that the media play their part in ensuring all of humanity can witness those first historic footsteps on Mars, and that people on Earth know the story behind those who are lucky enough to represent our species on another planet.

But if I'm honest, completing this particular entry will probably elicit the same muted response I had after being featured in the Guardian or the Huffington post, and much the same as appearing on Al Jazeera and CNN too: "Oh that's nice, at least they're interested". Because I didn't sign up for a one-way mission so I could *talk* about it - I signed up so I could *move to another fucking planet*.

For the last few years, my inner introvert has been fighting progressively harder against every bit of media exposure I've agreed to do. There have been some genuine media highlights in the eight years since I first discovered Mars One, especially meeting some of my science & journalist heroes like Andy Park and Dr Graham Phillips. But those positive experiences with the fourth estate have been more than outweighed by the negative ones, and most of the time I've been on air it's merely to answer the same questions coming from different faces.

Back in 2017, I got so sick of being asked questions in interviews that a glance at Mars One's website or Wikipedia could have instantly answered, I decided to publish *Becoming Martian* so I could throw it at people too lazy to use Google.

"How long does it take to get to Mars?"
"About seven months, but it depends on a lot of things - you should read my book."
"How will you protect yourself from radiation?"
"Dirt - it's explained in the book."
"How do you go to the toilet in space?"
"Trust me, it's explained in unnecessarily graphic detail... IN THE BOOK"

But as soon as I published a book answering everyone's science questions about humans on Mars, the focus immediately shifted to *me*. Suddenly everyone wanted to know *why* I signed up, how it's impacted my life, my family, friends and relationships. Which I'm sure we can all agree is *precisely* what every introvert trying to escape humanity desperately wants - more fucking attention. Which is why you're now reading a book written expressly to avoid any more interviews where I'm asked awkward personal questions like "What does your girlfriend think?", or having to deal with the fall-out from answering "Which one?".

So while I'll probably complete this entry regardless, my only real interest in having my stupid face splashed across the front of a newspaper is as a side-effect of contributing to something truly remarkable. To have a genuinely species-changing effect, any mission to Mars will need to share the personal stories of those representing humanity on another planet, as they'll blaze a trail for others to eventually explore even further. But if I can somehow avoid answering "What if you meet the love of your life before you go?!" for the 4739th time, that'd be even better.

67 – Visit... [Earth]
This entry has a total of twelve places to visit, and so far I've ticked off five of them: Uluru (Australia), Sydney Harbour (Australia), the Great Barrier Reef (Australia), the Pyramids at Giza (Egypt), and the Golden Gate Bridge (San Francisco). Which means I still need to see the Grand Canyon (Arizona), the Colosseum (Rome), Machu Picchu (Peru), Christ the Redeemer (Rio De Janeiro), the Great Wall of China, Angkor Wat (Cambodia), and the Taj Mahal (India).

The Grand Canyon is one of those things I've been *close* to multiple times, but I've never committed the extra few days in the US to see it. What I'd love to do is hire a campervan and see it on my way to Las Vegas to "Win Big at a Casino", then to be abducted/probed near Area 51 as I drive along the Extraterrestrial Highway. The Colosseum is relatively easy if I'm heading to Pompeii and Stromboli to feed my volcano fantasies. Also if I'm heading to South America to walk the Inca Trail to Machu Picchu, then it won't take much to divert briefly to Rio so I can see the 700-ton soapstone Jebus as well.

The last three just aren't on my immediate radar though. I'm sure it'd be incredible to see the stunning architecture of the Taj Mahal and Angkor Wat. Still, I don't care enough to visit India and Cambodia just for a couple of old, pretty buildings. Likewise, it *would* be cool to see something as iconic as the Great Wall, but it's not a strong enough desire to organise a holiday in China. Making a trip *just* to visit the Taj Mahal, Angkor Wat, and the Great Wall isn't particularly appealing either - it'd probably feel a lot like stamp collecting. Maybe other travel opportunities will take me to the same neighbourhoods, and I'll rave about the experience afterwards. But right now, I'm struggling to direct much energy or attention towards completing this particular entry.

72 – Have Enough Money to Do All the Things on This List [???]

"Having enough money" has always been the loosest entry on the entire list, and it's become even harder to complete ever since I signed up to live in a socialist Martian dystopia. The book itself says "you'll need an estimated amount of ten million", but it was published in 2004 and never actually defines if that's British Stirling, US Greenback, or some other currency. If it's Indonesian Rupiah, then I'm sorted because that's only about USD$700. Given a Zero-G flight is USD$5000 on its own though, I sadly suspect I'll need to start saving those Yankee pennies.

Rather than trying to go through all the remaining entries and estimate how much money I'd need to complete them, Elon Musk has provided a much simpler and clearer target for me to aim for: USD $1 million. While I'm entirely committed to sticking with Mars One as far as it'll take me, Elon wants to eventually offer people the opportunity just to *buy* a ticket to the red planet. SpaceX is optimistically aiming to put the first humans on Mars in 2024, who'll be there primarily to set up a rocket fuel plant to manufacture the methane they'll need to return to Earth. Elon reckons there's a 50-50 chance the first crew will all die on the way, but I'd still volunteer on a 50-50. After all, it's precisely the same odds Neil Armstrong gave the first lunar landing, and he was the one commanding it.

After that initial mission to build the methane plant, SpaceX plans to refit their spacecraft to carry 100 people to Mars at a time for roughly USD$500,000 per person. The cost per person will drop as they build more rockets and the economics of scale improve, and their goal is to eventually offer a round-trip to Mars for about the same as the US median house price: USD$250,000. So Austin Powers jokes aside, why am I aiming for *$1 million*?

That first mission to set up the fuel plant is damn dangerous, but at an average cost of USD$3 million per astronaut, it'll also be cheap enough for the likes of NASA to buy the entire thing to ensure American boots land on Mars first. So if I'm going to be on one of the later rockets carrying 100 people to Mars, I'll only need *half* a million dollars for a ticket. Unfortunately, I may need to pay for *two* seats because of an incredibly nerdy bet I made with a space geek of an ex-girlfriend.

Lisa rather cynically believes that NASA's "Space Launch System" rocket will be cancelled after it's first launch. I'm even less optimistic though, and don't think it'll *ever* launch at all. So whoever is wrong has to buy tickets for both of us to Mars. Which means if NASA's ludicrously delayed and obscenely over-budget rocket *does* ever get into space, even once... then I'll need USD$1 million to cover my wager with Lisa. If I'm right and SLS never launches though, then *she* has to pay for our seats instead... and I'll use my money to pay for another two space nerds to fly with us.

73 – Stand on the International Date Line [Earth]
This particular 101 things entry initially seems straight forward, but it gets progressively more complicated the more you look into it. It used to simply be a case of flying to Fiji, making your way to Taveuni then standing on the line. The 180-degree meridian runs directly through the island, so for years it had two time zones with a marker where you could quite literally jump between today and tomorrow. The Fijian government thought it was a bit ridiculous though, so they made it someone else's problem by pushing the International Date Line into the ocean between American Samoa and Independent Samoa.

The old dateline marker is still on Taveuni though, so you *can* bounce around it if you want to. You're not technically changing time zones when you do though - you're just making yourself look like a tool. Even if you caught the ferry between American Samoa and Independent Samoa, you still wouldn't be "standing" on the International Date Line unless you were 1.5 kilometres underwater. Since that's 80 times deeper than your average scuba diver is qualified for, and twice the crush depth of a modern nuclear submarine, "standing on the date line" probably isn't a viable option anymore.

I've crossed the dateline plenty of times flying from Sydney to LA and back, but if flying over it doesn't count then neither should taking the ferry. The 180-degree meridian also crosses Antarctica, but no one draws the International Date Line down there either. With months of daylight followed by months of darkness, "time" in Antarctica takes on a very different feeling, so everyone there simply uses UTC or a timezone from the country responsible for their base.

There's no clear way to complete this entry any more, but I'm kind of okay with that. If I'm ever in Fiji, I might dance naked around the dateline marker and get arrested. Or if I'm in Antarctica and near the 180-degree line, I'll dance around naked and get hypothermia instead. But this is one of the 101 entries I won't feel bad about leaving incomplete when I strap into a capsule bound for Mars.

82 – Build Your Own House [Mars]
If I'm honest, the chances of me owning or building a home on Earth are slim at best. For years I've been a firm believer that if you can't carry it, then you can't keep it. Which is why my house has been my backpack for the last decade, and why I get extra avocado at every brunch.

As I've shuffled through my 30's like some kind of deranged space hobo, I've also witnessed more of my friends suddenly start babbling about "settling down in the suburbs" with the "love of their life" to "have a few kids". And then they all get offended when I look at them in wide-eyed horror and ask if they can smell toast, because they're clearly having a stroke.

I'm not opposed to eventually scraping together a little timber for a remote Unabomber-style log cabin, maybe even making a treehouse and turning into a fully-fledged Ewok. But I've been moving around the Earth for so long it's tough to imagine building anything more than a temporary squat while I'm working on my next manifesto.

That said, I *will* eventually build my own house… but it'll be on another planet. While the first Martian habitats will be self-contained landing modules and inflatable units fabricated and launched from Earth, they'll still need to be connected by crews on the surface of Mars. As we strive to become independent from Earth, we'll also start manufacturing bricks and glass out of Martian dirt to build homes without relying on supplies sent by you filthy Earthlings. So even if I don't create an Unabomber cabin or an Ewok-treehouse while I'm here on Earth, I'll eventually build my own Martian home from the bricks up. I'm just grateful I'll be a very long way from your suburban family nightmare when I do.

85 – Visit the World's Tallest Buildings [Earth]
Since "101 Things To Do Before You Die" was first published, this entry has changed considerably. Different skyscrapers have been cancelled, expanded and changed as everyone races to build bigger dick-substitutes into the sky. But of the original 16 on the list, there are still 14 to see. Paul and I saw the Empire State Building and the Sears Tower during that same 2005 trip when he gave me the book.

However, it wasn't until early 2017 that I ticked off a third with the Petronas Towers. And *that* only happened because I was visiting Malaysia to make the terrible/hilarious mistake of trying to get back together with an ex-girlfriend.

The truth is I just don't care about tall buildings. Of the remaining 11 left on the list, four of them could be ticked off by a single visit to Hong Kong, another two in Shanghai, and two more in Shenzhen and Guangzhou. But going to any of these places just to tick buildings off a list would once again feel like stamp collecting. Neither Taipei 101 or One World Trade Center (the "Freedom Tower") have ever held much appeal either - if you want an incredible view from above, then how could a building ever beat the view from low Earth orbit?

The only building on the list that has ever held even a passing interest also happens to be the tallest, the Burj Khalifa. It's not that I care about telling people "I visited the tallest building in the world" though. It's far more about seeing where Tom Cruise ran around the outside of this ludicrous building for a scene in Mission Impossible 4, and wishing I could try some of those high-wire antics too.

88 – Get Married Unusually [Earth or Mars]
Even before the possibility of life on Mars, getting married has never really been on my to-do list. There has only ever been one time that I've thought legitimately about popping the question to someone, and it was in the misguided hope that getting engaged might save what was in retrospect a truly terrible relationship. And when I say "terrible relationships", I mean being stuck with Sam Neill on a spaceship that weeps blood would have been easier than continuing to share an apartment.

These days if I hear a little voice in my head whisper "Maybe we can rescue this by getting engaged", it acts like a screaming klaxon alarm that I'm in a haunted relationship or spaceship, and I need to find an escape pod immediately.

A Martian wedding could be an option though, or even one in space because you're bored and trying to kill time. The honeymoon is going to suck, but saying "Main Engine Start" before saying "I do" would certainly count as an *unusual* wedding to most Earthlings. There seems to be plenty of folks who have already evolved past the quaint concept of single-planet matrimony too, because for a guy loudly planning to leave humanity for another planet I've received a surprising number of proposals. I'll admit there might be other factors at play here though, because everyone who's popped *me* the question has been an educated American woman, and none of them were interested in getting married to an Australian *before* an angry orange manatee became their president.

Technically I've already been engaged once, but it was to another comedian after she called me chicken and dared me to propose. Both of us were in the US at the time and joked about eloping in Vegas, but it was never an earnest consideration. As lazy and emotionally-unavailable comedians, neither of us cared enough to even be in the same US state at the same time, let alone going through the hassle of actually meeting up to be married by a fat Elvis impersonator.

If I *did* meet someone incredible on Earth and we had a relationship based on something more than comedy, a schlocky 90's space-horror, or escaping Cheeto Hitler, then I guess getting married on Earth could be an option. We'd have to modify the marriage liturgy to "Till Mars or Death do us part" though because I am not getting married if I have to stay on this planet.

So far the stars haven't aligned, but if I do meet someone who's up for all of that, then rest assured I'll do my best to make the wedding super weird for everyone involved.

90 – Join the 16-mile-high club [Earth or Mars]

When I first discovered Mars One in September 2012, I immediately knew it was the most obvious path I could pursue to space. But it took another six weeks before I was sure I'd give up absolutely anything else to make it happen. I'd spent the morning watching a live-stream of some Austrian guy as he sat in a pressure suit and slowly ascended into the stratosphere under a balloon. While it was certainly inspiring to see someone set an altitude record, for me the real moment of truth came when he opened the capsule door, and the tiny amount of warm air inside suddenly puffed out into the empty black sky.

I've seen the same phenomena while parachuting: the lead jumper opens the aircraft's side-door, and the warm air inside the cabin instantly puffs out as the cold air outside rushes in. That blast of cold air serves as a chilly reminder that things are getting serious, and everyone should do their final gear checks before they jump out of a perfectly good aircraft. Within seconds you're at the door - looking to the horizon, with the sky above and the Earth 14,000 feet below. And then you leap into the emptiness.

But the camera over this Austrian guy's left shoulder wasn't looking at the horizon from 14,000 feet. It was at 128,000 feet, and that empty sky wasn't blue - it was black as the night. Another small shower of ice crystals dropped from a boom above the capsule door as he opened it. The Austrian switched to his suit's internal oxygen supply, took a moment to calm himself before unbuckling the seatbelt, slid to the open edge and stood on the capsule's porch as it bobbed gently in the stratosphere.

He saluted the sky, said "I wish you could see what I can see. Sometimes you have to be up really high to see how small you are. I'm going home now." And then he jumped into that empty black sky.

On October 14th, 2012 Felix Baumgartner became the first human being to break the sound barrier without propulsion, reaching over 1,300 kilometres an hour and breaking three world records with a freefall from 39 kilometres up. Crouched over a laptop in a friend's lounge room, I watched it live with my girlfriend at the time. And as Baumgartner landed and kissed the Earth, she turned to me and said: "I understand now why you need to do this".

Watching that record-breaking jump so soon after discovering Mars One was an extraordinary and defining experience. But the truth is I didn't care about much that happened after the capsule door opened. That brief puff of warm air instantly freezing against the black sky told me all I needed to know - that one day I'll leave the atmosphere behind to see Earth's sky turn black as well. But instead of looking down and saying "I'm going home now", I'll be looking towards the darkness and whispering "Fuck off, Earth" as the rocket's second stage kicks in.

94 – Get Something Named After You [Mars]
Much like getting my stupid face on the front of a national newspaper, I don't care about having something named after me - my priority is seeing humans living on another planet. It doesn't matter if that's me or not - it's just important that it happens. I'm also not here to inspire you or your kids, but if striving to be one of the first people on another planet has that side effect, then so be it.

Getting something named in my honour is really out of my control anyway. However, I'm particularly determined to avoid having a high school named after the first Australian leprechaun on Mars, mostly because my surname is Richards aka "Dick" with an "S". If I do wind up on the first crew to Mars, it's almost inevitable though. So while I don't care about legacy and will not encourage anyone to name anything after me, I will send a pre-emptive congratulations to the 2035 graduating class of Josh Dicks High. Aim for the stars, kids!

99 – Confess [Earth]
The only class I ever failed in high school was religious education, and the last time I walked into a church was while I was going through basic training with the Australian Army in 2004. We'd been ordered to attend church on our first Sunday, and the roof legitimately creaked as I stepped over the threshold. So it's unlikely I'll be threatening the structural integrity of a Catholic confessional any time soon.

That said, I am a strong proponent for being open and honest about everything, provided the details won't deliberately hurt someone else. Besides being a way to avoid awkward personal questions during interviews, the underlying concept to this book is to be "radically vulnerable" and share every aspect of my experience as someone who's potentially going to be one of the first human beings to walk on another planet. In its way, this book *is* my confession - opening the closet to reveal anything even resembling a skeleton, that way no one can claim later I tried to hide it. So be prepared for some serious overshare. Some folks wait their whole life before confessing on their deathbed, but weirdly enough you're reading my confession right now, and I'd encourage you all to do something similar. Let go of the guilt, treat every day like your last, and don't wait until your last breath to tell people how you truly feel - tell them to get fucked today instead.

100 – Reach 100 Years of Age [Mars]

If surviving to 30 was a bit of a stretch, then making it to 100 is a serious ask, even if we *do* ignore the fact I'm planning to be launched by thousands of tons of explosives at a cold desolate planet, baked with cosmic radiation on the way there, then baked again whenever I step outside my underground Martian lair. One Martian year is also nearly twice as long as a year on Earth, so it'll be a long time between birthday cakes. However, if we stick with counting "Earth years" then there's probably a higher chance I'll make it to 100 on Mars than I would if I stayed on Earth! Provided of course that we don't explode during launch, get hit by a massive solar flare in transit, land on Mars at "cratering" velocity, or die in one of a thousand other ways trying to get from the surface of Earth to our hobbit hole on Mars. Because once we're safely on the red planet, we're essentially living in a permanent, low-g health spa. The gravity on Mars is 38% of Earth's, so our hearts don't have to work as hard to pump blood, and there's less daily wear and tear on our bones and muscles as well.

There's still no shortage of things trying to kill us on the way there, and once we arrive we probably don't want to get sick either - the nearest hospital is a bit of a haul. Also, don't breathe the Martian dust, unless you like saying "I love the smell of perchlorates in the morning" and hate having a functioning thyroid. Between the reduced gravity and the close monitoring of the food, air and water in the habitat though; the folks on Mars will be living in an environment that'll give them a better chance of reaching 100 than most people on Earth have. Although if I do make it to 100 on Mars, it's not like I'd get a telegram from the Queen anyway, because these days she just sends an email! I don't even know how you'd get a telegram on Earth anymore anyway, but I do know the only email that I'll wait around on Earth to read is the one that tells me when I'm leaving this rock.

Chapter 3 - A Real Bucket List

There's no point doing all these things from someone's list if they're not what you genuinely want to experience though. There's no denying Richard Horne's "101 Things To Do Before You Die" has been an incredible guiding light for me through the last 15 years, serving as both a source of inspiration and as an actual list to work through. If nothing else, entry #71 "Have Adventurous Sex" has provided the fodder for some fantastically weird adventures.

However, there's been one uncomfortable question that's become progressively harder to ignore each time I've completed another of the 101 Things - what's next? And now with just 29 entries left, I'm finally starting to admit that I've done pretty much everything on the list that I ever wanted to do.

There are obvious exceptions, of course. Seeing the aurora and finally watching a rocket launch are two entries I'm quite actively trying to complete, because I'm confident they'll be utterly life-changing experiences. But if you had 10 to 15 years before blasting off to Mars, would *you* spend your days trying to see the world's tallest buildings? And if you're spending seven months in microgravity on the way to Mars, would you be disappointed about missing a zero-g flight before you left?

Signing up for a one-way mission to Mars and choosing to die on another planet tends to fuck up most people's bucket lists. Still, it's easy to put off things you genuinely care about when you assume you've got 15 years to work with - it's simply too long for many people to properly contemplate.

So what if you had 5 or 10 years instead? Why not dial up the pressure even more and ask how you'd live each day if you only had *one* year left? It doesn't matter if it's a one-way trip to Mars, a car crash, a stroke, or some sort of horrifying brain parasite - if you knew you had one last birthday, one last Summer, one last Christmas... what would you do?

Personally, there are a few Olympic sports I've always wanted to have a half-assed crack at trying. I grew up shooting, competing in pistol and small-bore rifle events through my teens and early 20's. Whenever the Olympics rolled around though, it wasn't the pistol or rifle shooting I stuck to the screen watching - it was the clay target double trap. The moving and pop-up targets the Army challenged me with cultivated an interest in fast *and* precise shooting, so the rapid and unforgiving nature of trapshooting appealed to that same side of me. The only reason I can give for never trying it is that I never knew anyone who went trap shooting, and until I started writing this book had never investigated giving it a go!

Switching from Summer to Winter Olympics, I've also wanted to try biathlon. Watching it isn't hugely exciting, but I've always wanted to try skiing some ridiculous distance with a rifle strapped to my back, stopping to pop off shots, then skiing off again to shoot at something else. It'd be hard as hell, yet there's something about that specific mix of physical exertion blended with self-control that has always appealed.

Just like every 90's kid who watched "Cool Runnings", I also figured hurtling down an icy slide in a half-ton bobsleigh would be pretty awesome. That is until I saw some other people in lycra throwing themselves feet-first onto a tiny one-person sled down the same track, and my eyes lit up like an ice-addicted maniac. However, any interest in luge was only fleeting the moment I discovered *skeleton*.

Skeleton is pretty much the same thing as luge, except instead of skimming feet-first just centimetres above the track at 100km/hr, you do it *head-first* instead. Australia isn't exactly renowned for snow and ice-based sports, so I do have some justification for not trying biathlon or skeleton just yet. There are no skeleton tracks on Mars though, so I'll have to treat myself to a Christmas in Norway before I leave. Nothing says "Christmas Spirit" quite like skiing cross-country to shoot things, before diving head-first down an icy death slide.

Sticking with a theme of trying to die while freezing, I've also looked quite seriously into swimming the English Channel. You need to book at least a year in advance, and prove you can swim for at least 6 hours in freezing water before an attempt. It's also bloody expensive paying for an escort boat to follow you across. Even so, there's certainly something appealing about stripping down to budgie-smugglers, greasing myself up with goose fat and then trying not to drown for 35 kilometres. Because apparently, I'm a reincarnated penguin.

If that wasn't enough self-inflicted punishment, I could always try slogging through 251 kilometres across the Moroccan desert for five days as part of the Marathon De Sables. It'd be one hell of a challenge, but I'm kind of hoping the English Channel will be enough for my masochistic streak. Redheads and sunlight don't mix well at the best of times, so running through sand in 50-degree heat for a week while carrying my food and shitting into plastic bags sounds unnecessarily unpleasant. Even for someone trying to die on Mars.

I'm certainly not opposed to covering ridiculous distances while carrying my house on my back though, because I've long dreamed of spending a few months on a self-supported "long walk".

Plenty of folks in the US spend months walking 3500 kilometres along the Appalachian Trail each year, and the 11,000 kilometre Great Western Loop would be an extraordinary challenge too. But if the black bears and ticks don't get me, then US Immigration is guaranteed to grab me the moment I exceed a 90-day tourist visa. Further south, there's the 3035 kilometre Greater Patagonian trail too. But if I'm going to visit South America, I might limit my walking to 3-4 days on the relatively touristy Inca Trail so I can finally see Machu Picchu. No matter how long I spend on DuoLingo, I doubt my Spanish will ever be polished enough to ask a llama farmer if I can sleep in their barn when I inevitably get lost on some godforsaken dirt track in the Andes.

Thankfully if I want to spend months walking through a picturesque wilderness, I only have to cross the Tasman Sea to a country that speaks something vaguely resembling English, where the greatest threat to my safety is overzealous Hobbit fans. The Te Araroa trail runs the full length of New Zealand's North and South Islands, stretching 3000 kilometres from Cape Reinga in the North to Bluff in the South, and takes about 4 or 5 months to walk the entire thing. The weather in the South island gets especially ugly between April and September, and there are still plenty of road closures to navigate well into October. Starting between December and January is supposed to be perfect though, so next Christmas if I'm not trying to become a bloody smear on a skeleton track in Norway, I might try hunting and eating hobbits while walking the length of New Zealand.

The appeal of walking alone for months has a lot less to do with admiring the beauty of Mother Nature, and far more to do with me being away from other human beings for a while. It probably seems like an odd thing for someone who's spent most of the last decade standing in front of crowds to say, but anyone who's helped me give a public presentation knows I'm quite happy being physically separated by a stage. And anyone unfortunate enough to see me do standup will testify that I don't particularly like being near the quivering bags of organs that pass for audiences in most comedy clubs either.

Live alone in a power-less wooden cabin while writing a 35000-word manifesto about the breakdown of civilisation? Sounds like bliss, which probably explains why I'm so keen to jump in a spaceship with three others and fly 20 light-minutes away from the remaining 99.99999994% of this species. It's also why I've always wanted to participate in a silent meditation retreat, considered studying Zen in secluded forest monasteries a bundle of different times, and continue my search for a job as the only caretaker for a haunted lighthouse. I'd probably stop short of staying for months in a hotel built on an Indian burial ground, but if anyone has any over-winter jobs at an Antarctic research station that's recently dug something alien out of the ice, then sign me up!

There *are* a few places I've dreamed about seeing that I've deliberately held back on visiting though. There's no doubt Johnson Space Centre near Houston and Star City near Moscow would be interesting, but I suspect I'd also leave with a mild sense of frustration at being there as a tourist rather than to train as an astronaut. My barely concealed *Contact* nerd would also love to see the Very Large Array (VLA) and Arecibo radio telescopes up close. However, I also suspect I'd leave New Mexico and Puerto Rico with a mild sense of disappointment that I didn't meet ET in the process.

Deep down, I know I don't *want* to work at these places full-time, and the best way I can help put humans on another planet is to keep writing & talking about Mars on my own terms, instead of disappearing into some enormous space science institution.

I'd probably still leap at the opportunity to visit Johnson, Star City, Arecibo or the VLA *if* I happened to be nearby. However that potential frustration and disappointment at seeing these places as a *tourist* and not as a scientist-astronaut means I'll probably never make deliberate plans to get my space nerd on, unless of course it's to see a massive rocket launch out of Cape Canaveral.

All of these different experiences would be incredible. Although many of them still feel a bit like taste-testers, as if I'm trying a bundle of different things to experience them and tell myself "Okay I tried that, I don't have to wonder about it anymore". There's no shame in experiencing all the weird and wonderful things that take your fancy, even if you have no desire to experience them again. Still, if I've only got a year to work with, I'm not going to waste much time with existential finger-food. The real focus has to be on things I've *always* dreamed of doing, and the most natural place to start is with things I obsessed over as a kid but then forgot about when the hormones hit in my teens.

The moment I heard on the 6 pm news that Andy Thomas had been selected as Australia's first professional astronaut, I knew I wanted to be an astronaut too. But within minutes I'd already convinced myself it'd be impossible. Even at the age of seven, I knew you had to be a US citizen to join NASA, and it felt wrong that to fly in space on the space shuttle I'd have to change my citizenship and carry the US flag on the shoulder of my spacesuit. My folks mentioned that Russia also sent people to space, and the Europeans occasionally did too. That's not what my seven-year-old eyes were seeing - they were seeing my role model as an Australian who'd become an American so he could fly in space, and I didn't believe I could ever follow in his footsteps.

Eighteen years later, I'd just left the military for good and was having a pretty miserable time trying to work out what to do with myself. It was just before Christmas 2010, and my housemates and I were having a few cheerful drinks in our dingy, mould-infested apartment in one of London's least reputable suburbs. One of my housemates considered herself a composer and had developed a reputation for playing the weirdest shit she could find on the stereo.

So when she suddenly asked: "Have you ever heard of the band Lemon Jelly?" I immediately rolled my eyes and groaned "No - do they drop jelly on keyboards or something?". She laughed and moved to the stereo as I held my breath, now expecting to spend the next 23 minutes suffering through the sound of Theremin music dubbed over an underwater recording of fifteen dolphins having a gangbang.

When the opening track to "Lost Horizons" started though, I was genuinely surprised to hear both a beat *and* a melody! Once it became clear the music *wasn't* going to abruptly change to a recording of Hitler opening the 1936 Berlin Olympics played backwards through a baby monitor, we all relaxed and returned to drinking cheap cider to dull the reality of our direction-less existence. Which is why when the second track faded in, it caught me completely off guard.

Within seconds I was entirely transfixed to the stereo. Recordings of the original Mercury, Gemini and Apollo radio transmissions played, spliced together with a gentle piano melody to create a short story about an astronaut encountering a ball of light in space. Just as the light envelops the astronaut, a drumbeat starts... and drunk in that filthy little London apartment; I burst into tears. At that moment nearly two decades stripped away, and I completely disintegrated - suddenly remembering that 7-year-old kid who just wanted to explore the wonders of this universe. The same 7-year-old Aussie kid who thought everyone should be able to go to space no matter which flag was on their shoulder.

I'm not ashamed to say that since that night, finding a way off this rock has consumed my every waking moment. There have been plenty of distractions and dead-ends along the way, but no matter what has happened, I've consistently come back to ask how I can contribute to humanity by leaving our home planet behind.
I've poured myself into studying astronomy, astrophysics, space engineering, space medicine & psychology, and they've all offered incredible potential careers along the way. Yet I know deep down nothing I might achieve in any of those fields while on this planet will ever come close to that all-consuming desire to be part of a crew that puts humans beyond Earth's gravity-well forever.

When I ask myself if there's *anything* I've been as consumed with besides the prospect of leaving Earth, there's only one thing that has ever come close. It's something I've obsessed with for almost as long as space, first started doing when I was 12, and continue to hold some ridiculous qualifications in it. And like space, it also allows you to explore the unknown. But it *isn't* masturbation. It's scuba diving.

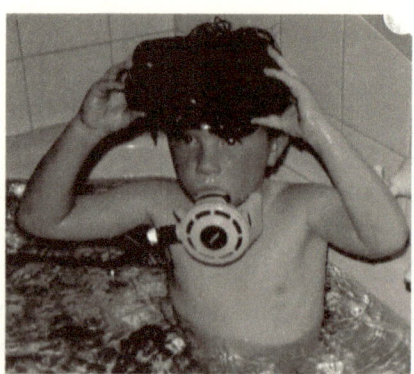

Age 7 - Lower edge of image cropped for your sake & mine

54 | *A Real Bucket List*

As a kid I grew up snorkelling in the Red Sea, ducking down with a camera to take photos of the clownfish and annoy Dad while he was training people to scuba dive - all the time wishing I could breathe underwater too. As soon as I was old enough to do my PADI Junior Open Water course, I had a cylinder strapped to my back and was sent under a reef to catch crayfish with Dad.

That early interest in scuba started to fade after high school though, as university and the Army took over. It resurfaced[1] while I was leaving the mining industry, desperately looking to do something I genuinely cared about instead of blowing huge holes in the ground for 12 hours a day. In less than a year, I became a PADI Master Scuba Diver Trainer, an SSI Open Water Instructor, and a TDI Advanced Trimix Diver.

Working as an instructor and training people to scuba dive *was* fun initially, but it also didn't feel like the underwater challenge I was after. Breathing from five cylinders filled with different gas mixes to explore a World War 2 era shipwreck in 90 meters of water was undoubtedly more challenging. But even deep trimix diving couldn't scratch the scuba itch I soon developed.

Because why *just* dive to ridiculous depths, when you could also dive deep inside an enormous underground abyss while breathing from a portable anaesthetic machine?

---

[1] "Resurfaced" is the only deliberate pun in this book, and I still threw up in my mouth a little writing it. A pun-loving ex-girlfriend used to tell me "A groan is as good as a laugh!" and to this day I have absolutely no regrets about breaking up with her.

Sometime around mid-2008, I came across an online article[2] from *Outside* magazine that changed my passion for diving forever. *"Raising The Dead"* told the story of an Australian cave diver named Dave Shaw, who during a record-breaking dive to the bottom of a South African sinkhole unexpectedly discovered the body of another diver who'd disappeared a decade earlier. Dave Shaw would die a few months later in the same sinkhole trying to recover that body from a depth of 270 meters, and his mentor Don Shirley would spend over 12 hours vomiting underwater after a helium bubble formed in his inner ear during decompression.

As odd as it might seem, reading about two people dying in an underwater cave while a third *nearly* died immediately generated an obsessive interest in cave diving and the closed-circuit rebreathers that make deep, long-duration dives possible. However, living in Perth through 2008 there was no obvious way for me to get involved with either - the nearest caves were more than a 1000km away, the nearest cave diving instructors were even further, and no one I knew was doing anything with rebreathers. So while I'd blow some very expensive bubbles and push my limits underwater diving deep wrecks off the coast of Western Australia for a few months of 2008 and 2009, my interest in diving faded again soon after.

A decade later though, and it's clear my interest in cave diving and rebreathers has never truly disappeared - it's just been on the backburner while I was being a boring adult for ten years. If I'm going to spend most of my remaining time on this planet diving into water-filled death traps though, then the sinkhole that Dave Shaw and Deon Dryer died in will *not* be the first entry on my "To Do" list of dive sites. But I'd certainly like to learn the basics of cave diving, and ideally by spending a month underwater in the Florida Springs.

---

[2] "Raising the Dead" www.outsideonline.com/1922711/raising-dead

After being trained not to drown in somewhere like Devils Den or Ginnie Springs, I can tick off another bucket list entry by swimming with a manatee, then head across to Cape Canaveral to finally see a rocket launch too! If cave diving lives up to expectations, then after my initial training in Florida I can look at diving in places like Mount Gambier in South Australia, the cenotes in the Riviera Maya on Mexico's Yucatan Peninsula, and maybe even that sinkhole in South Africa after all. But I'd still be happy to leave this planet if I only ever learned the basics of cave diving before I went - anything beyond that will be a bonus.

Just like cave diving, the desire to use a closed-circuit rebreather has remained lingering in the back of my mind ever since I read that Outside article about Dave Shaw's death. Standard scuba gear is "open circuit", so you inhale breathing gas from a cylinder, then blow it out as bubbles. However, the deeper you go, the more gas you exhale with each breath - a single breath at 90m is equivalent to 10 puffs on the surface.

Instead of breathing that gas out into the water though, a closed-circuit rebreather passes your exhaled breath through a "scrubber" to remove the carbon dioxide, replaces the tiny fraction of oxygen your body consumed, then loops the replenished gas back for you to inhale again. Because you're not using any gas besides the small squirts of oxygen your body metabolises, closed-circuit rebreathers give divers the ability to go much deeper and stay for far longer than is possible on regular open-circuit scuba. You've just got to afford a piece of diving equipment that's about the same price as a new car, and then keep this electronic underwater life-support system from feeding you the wrong percentage of oxygen and killing you.

There's no shortage of different closed-circuit rebreathers out there, and because they've spent an uncomfortable amount of money on diving gear everyone will argue that the one they use is the "best". From the "Mk 15.5" that Dave Shaw used, the "JJ-CCR" my first technical diving instructor swears by, to the "PRISM 2" my Dad has always wanted, to the AP Inspiration, the rEVO, the Sentinel... the list is endless. However, when I first started researching closed-circuit rebreathers in 2008, I seemed to zero in on the "Megalodon CCR" from Innerspace Systems.

I could go into all the technical details about how it's light, rugged, uses front-mounted counterlungs, has been used in caves to over 200 meters, blah blah blah. But the truth is I've never been able to fully explain what it is about the "Meg" that appeals to me - I just know it's the rebreather I've felt drawn to trying the most. Maybe with 10 or 15 years on Earth, I'll get right into rebreathers, try out a Megalodon sometime and decide I *don't* like the way it works after all. Or maybe I'd even build a rebreather myself as some folks do. But before I leave this planet, I'm determined to learn to cave dive AND how to use a closed-circuit rebreather... and that I'll never truly satisfy my curiosity until I try a Megalodon CCR at least once.

While cave diving and closed-circuit rebreathers are obviously at the very top of my diving wishlist, that doesn't mean there's not a ridiculously long list of *other* scuba diving adventures I'd also love to embark on before leaving Earth! During the armistice negotiations at the end of World War 1, most of the German Navy's "High Seas Fleet" wound up at the bottom of Scapa Flow near Scotland's Orkney Islands, after the German Admiral in charge ordered his fleet scuttled to prevent the ships from being seized by the Allies. Today, divers from around the world flock to Orkney every summer to explore these historic wrecks, so I'm keen to look inside at least one of the three battleships that remain.

In a similar vein, Micronesia's Truk Lagoon was a significant Japanese naval base during World War 2 until the US spent three days turning it into the world's biggest ship graveyard. After Jacques Cousteau and his buddies explored it in 1969, Truk Lagoon quickly became a global beacon for wreck divers as well.

Having already dived on my fair share of wrecks though, it'd feel slightly morose to keep looking at the rusting past instead of experiencing the vivid colours of life right now. For all the heavy-duty deep wreck diving I've done, none of it ever struck me speechless the way snorkelling with a whale shark did - watching the world's biggest fish slowly emerge from the depths and swim straight up at me is not something I'll ever forget. After dashing out of the way of it's gaping maw and starting to swim alongside this 10-meter leviathan, it quickly became clear I wouldn't be keeping up. The gentle flicks of its gargantuan spotted tail could push 20,000 kilos of fish through the water much faster than my pathetic human legs could ever hope to shift 65 kilos of ginger weirdness, even at a sprint.

Breathing underwater might feel both extraordinary and perfectly natural to me, but doing it next to an incredibly powerful animal that spends its entire existence in water was a genuinely indescribable and awe-inspiring experience. I literally can't think of the words to adequately describe it. For sheer majesty, the only things I can imagine topping a swim with a whale shark would be scuba diving with a whale, or possibly night diving with the enormous bat-like Manta Rays at Garden Eel Cove in Hawaii.

You might also expect me to be the type who'd list "Diving with Great White Sharks" on my scuba bucket list, but a decent chunk of my teens were spent spearfishing in places with names like "Shark Bay".

Which, unsurprisingly, had more than a few Tiger, Bronze-Whaler and Hammerhead sharks swimming around. All of whom became less than friendly after some teenager had spilt fish blood in the water. I'm sure diving in a cage with Great Whites would be memorable, and I'd happily join a friend who wanted to see them up close. But thanks to those spearfishing trips with Dad, I already have plenty of shark-related stories to terrify people out of the ocean.

There's the time when I was 12 and had to hang off my Dad's back as he pushed away one shark with a speargun while another circled beneath us. Or when I was 15 and had to hide in a school of baitfish while half a dozen black-tip sharks circled outside. A year later, I had an angry hammerhead chase me 100m across the surface and back into the boat. And on the same trip I watched a 15 foot Tiger shark bite another shark in half - it was **awesome**. Although I didn't need to stick around to see if it was still hungry afterwards - when sharks start eating other sharks, you get out of the fucking water.

If I'm going to voluntarily get in the water with something big *and* vicious in the future, I'd much rather it be an Orca. At least that way if I dress up as a seal or penguin, they'll play with my lifeless body for a while instead of merely eating me as a shark would. However "Killer Whales" don't truly deserve their reputation for being dangerous to people. Sure, they'll team up to methodically murder seals and have been filming kicking penguins around for fun. However, the only time orcas have ever killed *people* is when *people* have locked a highly social creature - with a brain six times bigger than ours - in a tank by themselves. I've got zero interest in swimming with an animal that's trapped in solitary confinement, although I'll happily grab the folks who've been keeping orcas in tanks and throw *them* in.

My best bet for swimming in the wild with a pod of these majestic 6-ton seas wolves is to visit the island of Lofoten during the Norwegian winter, where the Orcas come to feed on the krill that congregate in the fjords. The orca show up a couple of months before the competitive snow sport season starts though, so I'll have to organise a separate trip to Norway once the Sun returns for my biathlon/skeleton death-slide adventure.

The other option for swimming with orcas might be near Antarctica. Besides being a lifelong dream to visit and dive the icy waters at the bottom of the world, there's also the added benefit of potentially being in the water with something else big and bitey that *does* have a record of murdering people in the wild for fun: Leopard Seals!

Steve (38) - Loves long swims, holding flippers, and the taste of penguin in the morning.

Everyone gets excited about seeing sea lions, emperor penguins and all that other adorable Antarctic "Happy Feet" bullshit. But you can keep all of that because I'll be busy swimming between icebergs with something that looks like a 3 meter long, half-ton mutant dinosaur that attacks people and punctures inflatable boats.

While they're aggressive, unpredictable and legitimately dangerous, there's still something appealing about being in the water with an intelligent and powerful predator that's genuinely curious about your presence. There's an extraordinary series of photos from *National Geographic* taken by photographer Paul Nicklen in 2006, when a female leopard seal he was in the water with, decided to teach him how to hunt penguins.

The leopard seal started by bringing Nicklen live penguins in her mouth and releasing them for *him* to catch. When the photographer didn't chase down and eat the wounded penguin she'd brought him, the leopard seal would pick it up again and drop it in front of Nicklen for him to try again. For four days this leopard seal kept trying to teach the photographer to hunt penguins - getting progressively more frustrated at his penguin-eating incompetence until she got pissed off enough to start killing the penguins outright and dumping their bodies on Nicklen's head. When he then appeared to reject her "gifts", the female leopard seal started biting the camera and blowing bubbles out her nose in a threat display to express just how pissed off she was that this weird creature wouldn't eat her penguins.

When I first saw the photo series in a 2008 *National Geographic*, I sat slack-jawed and awe-struck by the photos for a full hour, and diving with a leopard seal has been near the top of my all-time bucket list ever since. There might be a few other diving experiences that would be pretty magical too, like hanging out with regular seals in the crystal clear waters of Lake Baikal in Siberia. Although if I'm honest, all I genuinely care about is visiting Antarctica to have a half-ton aquatic murder machine throw flightless birds at me before I leave the planet.

One of the most common questions I get asked since signing up to Mars One is what I would miss from Earth, and while I'd bullshit people with nonsense about "not missing what you can't have", secretly I've worried that my honest answer might be scuba diving. The only thing that has ever come close to matching my desire to explore outer space is how I feel exploring "inner space", and we're unlikely to have leopard seals delivering dead penguins to us near the south pole of Mars anytime soon.

Even so, I avoided scuba diving for years - partly I was put off by the equipment cost, the logistics of getting it to exciting places to dive, and a mild sense that it wasn't *quite* the kind of exploration I sought. But the truth is I initially avoided scuba diving to avoid questioning my commitment to living on Mars. Travelling between Dublin and Boston in 2016 for a comedy tour however, I took the opportunity to stopover in Iceland on the way. In the process, I unexpectedly reconnected with my love of scuba diving, while simultaneously realising that I'd ultimately be okay giving it up forever to live on another planet.

Silfra is a freshwater fissure in Iceland's Thingvellir National Park, formed by the North American and Eurasian tectonic plates slowly pulling apart. With visibility exceeding 100m, it gets consistently placed in the world's top 10 greatest dive sites. While there's a cave system that technical divers can drop more than 60 meters into, the most beautiful parts of Silfra are just 5 to 10 meters below the surface. Rain running off nearby mountains is filtered through the limestone beneath Thingvellir valley for decades, so when that water emerges into the Silfra fissure, it's become the clearest freshwater on the planet. The water is so clean there's barely enough nutrients to support aquatic plants or fish, so it feels like a barren moonscape as you drift gently between lifeless black and grey volcanic rocks.

With water temperatures averaging between 2-4 degrees Celsius, only the mentalists jump in without a drysuit. But being in this freezing water without getting more than your face wet just adds to the other-worldly experience. I only spent 90 minutes in the water at Silfra, and I never went deeper than 12 meters, yet those two dives are without a doubt the most extraordinary scuba experiences I've had to date.

A Party Clown of the Floating World

When I look back on diving at Silfra though, I realise two things. Firstly, I was a little too enthusiastic jumping into 2-degree water, which only adds further evidence to the theory I'm a reincarnated penguin. Secondly, and more importantly, I have to admit that it may not be scuba diving itself that I love so much. What I suspect I'm obsessed with is that gentle sense of weightlessness and the awareness of breath as you carry your life support on your back. Diving seems to act as an interconnected moving meditation for me as I float slowly through an alien world I observe with the sound turned down, and the colours turned up.

When I ask myself why I prefer scuba diving in a dry suit, I realise it's not just about comfort - it's because the sensation of water creeping through a wetsuit breaks that open-eyed meditation. When I reflect on why closed-circuit rebreathers hold so much appeal considering traditional open-circuit scuba is so much simpler and cheaper, I realise it's not just about going deeper or staying longer. It's because, with every exhaled breath, the stream of bubbles reminds me I'm in water instead of space.

Until I experience it first hand, there's no way to be sure. But I strongly suspect that stepping out of an airlock in a Mars suit to set foot on an alien world with 38% Earth gravity will instantly trigger everything I love about scuba diving and so much more. Until then, whenever people ask what I want to do before leaving Earth, I'll just have to keep telling them I want to swim with a murderous leopard seal while it dumps dead penguins on my head.

Chapter 4 - Fear

It's all well and good to have two bucket lists full of things to do before leaving Earth, as well as a homicidal comedian threatening to end it if you chicken out and stay. But what if there's some reason I genuinely *can't* go? Or what if the rocket blows up on the way to Mars? What if the retrorockets on the capsule don't fire, and we slam into the red planet at terminal velocity? What if the entire project fails to get funded and everyone gives up? Or what if the whole mission to Mars *does* go ahead perfectly, but I'm never selected?

There are a million "What If?" questions that people ask me about this whole endeavour, and somehow they're even more tedious than a simple "Why?". What if Sarah gets bored and murders me *before* launch? What if a leopard seal decides to dump *my* head on a penguin instead of the other way around? I want to scream "Who the fuck cares?!", but the question I politely ask people in return is "What if we never try?". I mean, why bother venturing into the unknown and trying to discover more about the universe and ourselves in the first place? It'd be so much safer to stay on Earth and focus on how we'll get by each day here instead, right? It's not like Earth has had five extinction-level events that wiped out over 75% of all species each time or anything.

You might think going to Mars is risky, but I'm pretty confident staying on a single planet is a hell of a lot more hazardous for us as a species. Extinction-level events come in all shapes and sizes, but they seem to have a distinct preference for wiping out the dominant animals of the time to make way for another order to take over.

The Ordovician-Silurian event around 445 million years ago and the Late Devonian extinction 376-360 million years ago both managed to obliterate most species of coral and invertebrate shellfish. However, that mass extinction helped make way for the evolution of larger vertebrate fish. The Permian-Triassic extinction might have wiped out 96% of all marine species around 252 million years ago, but that's also precisely when the dinosaurs suddenly started kicking the shit out of the amphibians for not evolving out of the water fast enough. However, the dinosaurs didn't begin their reign of terror properly until the Triassic-Jurassic extinction knocked out most of their competitors around 200 million years ago.

The dinos spent the next 135 million years stomping on pretty much anything else, until about 65 million years ago when a giant fucking rock fell out of the sky over Central America and wiped out anything bigger than a wombat.

You will die, and Pebbles will take over.

It turns out we're actually in a sixth extinction-level event right now. It seems puny humans and their fancy opposable thumbs have been leading the Holocene extinction for the last 12,000 years, and we've been annihilating other species 10-100 times faster than any other extinction event in Earth's history. Wooo? Go team?

Now imagine if the dinosaurs hadn't wasted all their time eating plants and each other, stopped kicking the shit out of mammals for five whole minutes and had instead started a space program to escape that asteroid? I doubt anyone would miss the Gulf of Mexico if giant lizards were still running around murdering things, laying eggs the size of basketballs, and LAUNCHING THEMSELVES INTO SPACE! Sure, if the Chicxulub asteroid hadn't hit then the mammals wouldn't have taken over from the dinosaurs, and humans would never have evolved. I'd still prefer the infinitely better "Dinosaurs survived" timeline though, where imperial Velociraptors fly landspeeders across the moons of Jupiter and hunt down those rebel Protoceratops scum.

Bird nerds like to point out that there's strong evidence that many dinosaurs (including T.Rex and Velociraptor) had feathers, and that many flying dinosaur species probably survived the Chicxulub impact and evolved into modern birds. So my apologies, not *all* the dinosaurs were wiped out - just the good ones. Today the closest living relative of the mighty T.Rex is the domestic chicken, which is just great, I suppose. It's not like I'd *want to* live in the timeline where murderous 9-ton killing machines roam the solar system in giant T.Rex-size space ships. Or where lab-coat-wearing Velociraptors toil long into the night trying to unify quantum mechanics with general relativity while fuelled by coffee and Pterosaur pie.

No, instead *we* get to live in the lame reality where the descendants of T.Rex and Velociraptor are monumentally stupid creatures that peck mindlessly at the ground but taste great with 11 herbs and spices. All because of Earth's chance collision with a giant space rock. Sometimes, evolution sucks.

While we're not currently tracking any potentially dangerous asteroid impacts, there's also nothing we could do to stop an asteroid barely 50m wide if we spotted one, let alone deflect a 100-kilometre wide planet-killer. Clone Bruce Willis and launch all the nukes you want, you may as well sing campfire songs when the big one hits. It's not like asteroids are the only things lurking in space waiting to murder all of us either. You won't even have time to light the campfire if a nearby Wolf-Rayet star collapses because the gamma-ray burst will strip the ozone layer in seconds before completely irradiating the entire planet. At least your roasted marshmallows will be perfectly warm and gooey like your insides, and thankfully there'll be no one left to play a ukulele.

That's just a fraction of what could wipe out *all* life on Earth though, let alone all the fantastic disasters that might only annihilate humanity and allow hyper-intelligent dolphins to rule this planet with an iron flipper. A virulent superflu, being cooked/frozen by climate change, global agricultural collapse, or a super-intelligent AI that decides humans are a threat and unleashes an army of genetically-engineered acid-breathing lemurs? Honestly, the options are endless.

One of my former housemates, who I'll refer to as "Lily", passionately believed humanity would eventually be wiped out by ants. After millennia of abuse from kids kicking over their ant hills or burning them with magnifying glasses, one colony would finally decide to strike back and bite some fat kid on the toe.

He'd squeal and run away, but the ants would now have a taste for human blood. They'd start nipping kid's toes everywhere, drinking the blood of our youth to absorb their power and grow larger. The ants will subsist on simple toe biting until they'd grown to about cat-size, at which point they'll start to hunt down and consume unattended children. As adults started putting ant-shields around kindergartens, Lily's ever-growing ants will form packs to hunt the elderly instead. By the time the ants have swollen to the size of a Great Dane, their blood-thirst will drive them to start assaulting our cities in broad daylight.

We'll turn our armies on them, but to no avail - the formic acid of their land-prowling soldiers will eat through the armour of any tank, while their winged counterparts will control the skies with sexually frustrated rage. We'll drop nuclear bunker busters from space on their subterranean nests, but their radioactive Queens will only spew out larvae faster. The planet will be left desolated - the air filled with toxic nuclear ash, the oceans poisoned, and the land littered with unwatched DVD's of David Attenborough's "The Empire of the Ants". By the time the ants have grown to human size, they'll have learned to look like us too - sucking out our warm and wet insides to leave behind a fleshy skin suit they can then wear to infiltrate the last pockets of human resistance.

With all the humans gone and the ants still mad with bloodlust, Lily believed they'd then turn on each other, and the Million-Year Ant War would truly begin. Armies of car-sized ants would roam the entire planet for thousands of years, striking other colonies and eating their enemies to consume their power and grow ever larger still. Every time one colony wiped out all the others, a civil war would break out, and the enormous ants would immediately start eating each other again.

The ants would grow in size with each battle, but would slowly dwindle in number as only the blood of other monstrous ants could now sustain them. They'd rage on in relentless insecticide until only two continent-sized ants remained. Fueled by millennia of ant-on-ant slaughter and knowing there can only be one, the final battle to the death would begin. Raging on for another million years, one ant would eventually slip on a crumbly bit of Bhutan; the other would strike a fatal blow, and the fallen consumed to leave a single, planet-sized murder ant. This final gargantuan ant would then eat the Earth itself, before floating off through the darkness of space to terrorise and slowly consume the rest of the galaxy.

And Lily believed all of this... because she'd completely disassociated from reality. Honestly, I spent six months coming back to that house each night not knowing if she'd be making dandelion soup or proclaiming she was the Queen of France. On the plus side when I go to Mars with three other people crazy enough to come with me, I'll already have had plenty of first-hand experience deciding if my housemates are "quirky" crazy or "cook the cat" crazy.

At least Lily understood the ridiculous and tenuous nature of human existence. I doubt many people fully grasp how astronomically lucky our species is to be here, or even how improbable it is for the *universe* to exist the way it does in the first place. Changing just one of the universe's fundamental physical constants by a few percent would have prevented stars from ever forming carbon, and life as we know it wouldn't exist. Even in a universe capable of supporting carbon-based life, there's so much that could have annihilated all life on this planet well before we turned up, and something is *guaranteed* to wipe us out if we stay here too long.

My only hope is that we escape the cradle of Earth before the ants annihilate our species. And that if we *do* manage to leave before the ants get a taste for blood, that it'll be the majestically vicious orcas who are triumphant over those smug, douchebag dolphins in the post-human battle for Earth. Or maybe some mutant species of giraffe takes over. Who knows? If we've finally evolved into a genuinely space-based species, then who cares who's in charge down on Earth?

Maybe I'm willing to jump at something as risky as a one-way mission to Mars because I've realised how ridiculously lucky we all are to be here. For every one of us, a remarkable set of circumstances have had to align to be here as we are right now. The odds of your parents' meeting when they did, coupled with the odds that a specific series of events would eventually lead to your birth and upbringing are all astronomical.

Yet those odds are stacked on top of the chance events that lead to your grandparents' meeting. Likewise, their births rested on the chance decisions of thousands of previous generations before them too. I'm still not sure if the whole thing was just some generally bad idea, but as Richard Dawkins so eloquently puts it: "In the teeth of these stupefying odds it is you and I, in our ordinariness, that are here."

Wet & Dry

Which brings us to one of the questions I'm most commonly asked: "Aren't you afraid of dying on Mars?". When I laugh gently, smile, and then say no, some confuse my reaction for naivety. Others assume I'm trying to cultivate a "quietly macho" kind of persona. But the reality is I'm simply not fussed about dying on Mars. In the last 35 years, there are so many things that could have turned out fractionally differently and resulted in me not being here to write this at all, so it's not naivety or bravado you're hearing - it's gratitude.

But I don't believe that surviving a series of close calls is enough for someone to be truly at ease with their mortality. For that, I think you need to experience something that scares you even more than the thought of dying does. Now maybe I'm wired wrong, but after nearly kicking the proverbial bucket a few times I've recognised that it's not dying that scares me the most - it's letting other people down when they need help. And I'm not talking about the "Can I borrow your trailer to move house this weekend?" kind of help. I'm talking about the type of "help" where you whisper to yourself: "People I care about will be coming home in body bags if I fuck this up".

The UK's Royal Marine Commandos are arguably one of the toughest military units on the planet. With the longest basic infantry training course in the world, they're also the only unit on the planet that can (and regularly does) send freshly trained recruits straight from basic training directly into a warzone. Every other military force puts their already qualified soldiers through "pre-deployment training", where the skills they've learned previously are polished through 3-4 months of high-intensity exercises to prepare them for active operations in a warzone.

Instead, Royal Marine Commando "basic training" directly combines four months of initial training with four months of preparation for frontline war-fighting. More than 75% of recruits never complete the eight months, but in 2009 I decided I was up to the challenge and joined.

Proof that without a beard, I look like a child AND a serial killer

Besides being amphibious infantry with a wide variety of roles, the Royal Marines are also the UK's arctic warfare specialists. Each year, Royal Marines head to Norway to take part in "Exercise Cold Response" - a Norwegian-led military exercise where soldiers learn to survive in -30°C temperatures. For several weeks they run and ski through the snow as well as taking part in "Ice breaker drills" - skiing into a hole cut in the ice over a lake, then dragging yourself out of the freezing water with a pair of ski poles.

With the expectation you'll regularly operate in sub-zero environments, one of the first survival skills Royal Marines learn is the life-saving "Wet and Dry Routine". The concept is simple: whenever you head out into the field, you carry two uniforms.

One uniform is your "wet" gear, which gets covered in mud and drenched by rain as you slog through whatever horrendous circumstances you find yourself. Anytime you're on the move, you're wearing your "wet" gear. At the first opportunity to stop and sleep though, you immediately put up a shelter, strip naked, and pull your sleeping bag and dry uniform out of the waterproof satchel you've kept buried deep in your pack. After quickly towelling yourself off, you throw on your dry gear and climb into your sleeping bag as fast as you can to try and warm up. And if you're *incredibly* lucky, you might even snatch a few minutes of sleep.

Doesn't sound too bad, right? The "dry" part of the routine where you put on the dry uniform and get your head down in a sleeping bag *is* bloody fantastic. It's just doing the whole thing in reverse when you need to move again that's the issue. Your dry uniform needs to stay dry so it doesn't soak your sleeping bag, because that bag will save your life in the cold. So as soon as you get up, you need to strip out of your wonderfully warm and dry uniform, repack them with the sleeping bag in the waterproof satchel, then <u>put the "wet" gear back on again</u>. Moving around warmed your wet clothes a little, and you never stopped in one place for long if you were wearing wet clothing. But climbing out of a toasty warm sleeping bag to put on a soggy, muddy uniform when it's cold enough to freeze leather boots shouldn't be anyone's idea of fun.

My commando troop practised wet and dry routine for the first time about a month after we started at the Commando Training Centre in Lympstone. Being thrown into the muddy pool under the regain ropes, then being told to get changed as quickly as possible before camping out on a football pitch overnight was less than fun. However, the real test would be a three-day exercise slogging through the mud in Woodbury Common a week later.

The first two days we were running up and down hills, kept up through the night, and regularly dunked in rivers to complete the trifecta of being cold, wet and tired. Just after sunset on the final evening, we started marching out along a road to where the troop commander decided we'd camp for the night.

I say it was a "road", but it was little more than a mud track so slippery you'd sink up to your knee in it. After four years with the Australian Army I thought I'd seen some pretty shitty mud, but this was undoubtedly the worst. It took the 60 of us nearly an hour to slog the kilometre and a half to the campsite, with recruits hobbling along on rolled ankles and falling in the mud the entire way.

Once we'd clambered off the track and in among the cover of the pine trees, we immediately started setting up our shelters, but the training staff told us not to get too comfortable. Each of us needed to build our shelters, conduct wet and dry routine, plot a series of navigation points on our maps while inside our sleeping bags, and be packed up ready to start a navigation exercise in just over an hour that would likely run through the whole night. If we worked quickly enough, we might get a chance to eat or maybe even a few minutes of sleep.

I threw up the shelter as quickly as possible, changed into my dry gear, plotted the nav points on my map, helped the guy I was sharing a tarp with plot his too, ate some cold rations, then put my head down for a brief nap. I'd *just* closed my eyes when another recruit ran up, shook me and whispered: "Oz, come quick - there's something wrong with Joe".

As one of the few guys there with previous military experience, I inevitably became a bit of a go-to for the younger guys who were understandably terrified of the training staff.

Most of our training corporals had already done at least three tours of Afghanistan, one was training for the Special Boat Service, and another had spent a lot of time with one of the UK's more shadowy reconnaissance units. These were some incredibly tough and unforgiving men, and no one ever wanted to even look at the training staff the wrong way, unless for some reason they wanted to make this hideous experience even more unpleasant.

However, I wasn't particularly pleased about being the first port of call instead.

"ARE YOU FUCKING *KIDDING* ME?! I *just* got my head down! What the fuck is wrong with that useless prick now?!"
"I don't know, but he's not talking. Something's really wrong."

Joe was *not* my responsibility, and I was thoroughly sick of being a troop "fixer" just because I had a little more experience. But I got up anyway - changing back into my wet gear, packing my sleeping bag and dry gear away, and grumbling as I followed the other recruit back to where Joe's shelter should be. As we walked up, I realised Joe wasn't under any cover at all. He wasn't even in a sleeping bag - he was just laying on a roll mat under the stars. A bit confused and still pissed at being back in wet clothes, I tried putting on a friendly voice to ask: "What's going on, Joe?". No response. As I knelt next to him, I asked again but stopped myself mid-sentence - I was now close enough to see *exactly* what was wrong.

I looked at the recruit who'd brought me over to Joe, and calmly but firmly said: "You need to get the training staff". He started to quibble, scared of how the training staff might react to being disturbed, but I exploded: "GET THE FUCKING STAFF NOW!".

It's hard to describe how insidious hypothermia is to people who haven't experienced it. While you might initially feel cold and whiny, as your core body temperature starts to drop often your desire to do anything about it disappears too, so it quietly eats away at your will to live until it's too late. I'm not sure if it's the way doctors define it, but we learnt to break hypothermia into three stages. Stage One hypothermia only starts when numbness truly sets in - when you can't stop shivering even when you tense your muscles up, and your perspective starts narrowing so all you can think about is how cold it is.

During Stage Two, you get clumsy because you've lost the function in your fingers and feet as your body pools what little warm blood it has left, around the vital organs in your chest. Stage Two is also when you see people start to give up: they'd be so tired, clumsy and numb they'd stopped feeling cold and just lie down in a stupor. If they *do* stop moving, Stage Three soon follows - the point where they can't save themselves.

In Stage Three, the limbs become useless as all the blood is pooled in the chest and head, and they usually stop speaking. Stage Three cases were beyond the point where someone could save themselves, and would generally end with someone falling asleep and not waking up again. I've heard of a fourth stage where people become manic and delusional, believing it's warm and in a burst of energy manically strips off their remaining clothes. I've never seen a manic phase, though. To me, hypothermia has only ever appeared as a slow, creeping death that quietly but effectively lulls people into a sleep from which they never wake.

By the time I knelt next to him and yelled for the training staff, Joe had already stopped speaking and was starting to doze off to sleep. Rather than going through the ordeal of doing wet and dry routine, Joe figured it'd be easier to just stay in his wet clothes for the hour. Already tired and ready to give up, he'd simply rolled out his mat and laid under the stars rather than going to the effort of setting up a shelter. It was about 5°C, which wasn't that bad compared to the -19°C we'd experience during a blizzard on Dartmoor a few months later. The difference on Dartmoor is we were mostly dry, but here Joe's clothes were still soaking wet after he'd had to jump in a stream hours earlier.

My first instinct was to get him out of his wet gear before putting him in his sleeping bag, so I started trying to get his outer jacket off. I'd just pulled it off when the training staff arrived - one of the corporals took one look, immediately pushed me aside and started jamming Joe into his sleeping bag with the rest of his wet clothes still on. Once we zipped him inside the sleeping bag, three of the training staff and I picked him up as the fourth training corporal and the other recruit slid a plastic sheet underneath, and we lowered Joe back onto it. All six of us grabbed an edge of the sheet and stood up, lifting Joe in our makeshift stretcher.

And then we ran.

There was no way an ambulance could get down the mud track, so when the troop commander radioed the emergency call through, the medics in the ambulance had replied they'd be waiting right where the mud began. We just had to carry him the kilometre and a half to get there.

Before I'd started training with the Royal Marines, I'd worked exceptionally hard to develop my fitness as much as possible in preparation for it. Part of pushing myself harder and further than ever before involved developing a voice in my head that would scream at me when I was tired or ready to give up, and it sounded a *lot* like Gunnery Sergeant Hartman from *Full Metal Jacket*. "Pick it up, Richards! Stop being so fucking weak!" would come booming through my head when I thought my legs were going to give out on a particularly torturous run, or when I was gasping for breath during a swim.

I hadn't heard the voice much during training, because the corporals were loud and angry enough that I didn't need to add a "motivating" voice in my head. As we ran to the ambulance, most of the training staff fixated solely on getting Joe there as fast as possible, while the other recruit was just trying to keep up. But the angriest of all the training staff started speaking to Joe like an old friend as we ran. This corporal - who seemed to hate recruits, and had casually threatened to shove a bayonet in my ear a few weeks earlier - was now saying "You comfy in that bag?" and "Stay awake buddy" while calling Joe by his first name... which wasn't "Joe", in case you were still wondering. Hearing that training corporal's tone soften for the very first time was when I started to believe the guy we were carrying was probably going to die.

As we ran, my forearm started burning with the pain of holding the plastic sheet. There was no way I could adjust my grip without letting go though - my hand had seized on. I wanted to cry out to stop and change hands, but I knew we couldn't. As we ran carrying one of my friends through some of the shittiest mud I've ever seen towards an ambulance, that inner voice suddenly came back.

But this time, it sounded different to anything I'd ever heard before. Instead of being angry or threatening, it just started repeating itself in a calm but unforgiving tone:

"Don't trip. Don't let go. Don't trip. Don't let go."

Because I knew if I tripped in that shitty mud or tried to adjust my grip on the sheet, then Joe was probably going to die.

Eight minutes after we'd picked him up on the plastic sheet, the six of us slid Joe into the back of the waiting ambulance. The four training corporals all climbed in with him, and the angriest one helped peel my fingers from where they'd seized onto the plastic sheet. Then the doors closed, and the ambulance sped off to Lympstone hospital. Standing there watching the taillights disappear into the darkness, the troop commander turned to me and said: "We need to move everyone up here by the main road".

The three of us stumbled back down the mud track, the troop packed up, and we all moved back along the mud track to the main road where we camped among the pine trees. Without the training corporals, the navigation exercise couldn't go ahead, so the troop commander told everyone to bed down and get a decent night's sleep instead. As soon as he finished speaking to everyone, he turned to me and said: "Richards, make sure they all get to bed okay" - he'd be in his tent several hundred meters away sorting out the paperwork from Joe's ambulance ride.

With a weary smile I said "No worries sir" - he left for his tent, and I went around the troop making sure everyone was setting up their shelters, getting into their dry gear and settling into their bags to sleep. Which is when I found two more that had gone hypothermic.

They weren't quite as far gone as Joe had been, but they were close. Long after the rest of the troop had fallen silent, these two were still stumbling around trying to strip out of the wet clothes they'd put back on after we'd moved camp closer to the road. Seeing them struggle with that exhausted clumsiness, I knew they were well and truly into Stage Two and would slip into the helplessness of Stage Three if they didn't get dry and warm soon. The troop commander was in a tent several hundreds of meters away though - without a radio, I knew the thick pine forest would block any cry for help. I tried to rouse some of the other recruits, but none of the others would stir from their sleeping bags - we were all cold and bone-tired, and I had been the one tasked with making sure everyone got to bed. It was just these two left, and soon they were beyond clumsy and sluggish - they were sitting down and starting to doze off.

There's something very scary knowing there is no one coming to help - it's just you, and the decisions you make, that will determine whether your friends live or die. It didn't feel like I had any choice though - it was my job to help these two out of their clothes, into their dry ones, and to scream at them any time they stopped moving. So that's what I did: I screamed myself hoarse to keep them awake while pulling them out of their wet clothes, then pushing them into their dry gear and sleeping bags.

Two hours after someone shook me awake to help Joe, the last two recruits finally got in their sleeping bags and quickly fell asleep. I remember briefly worrying they might not wake up the next morning, but I also knew I'd done all I could by making sure they were wearing warm and dry clothes inside sleeping bags rated for the Arctic. However, as I wandered back to my shelter and reached for the zip on my jacket to remove my wet gear, I suddenly realised that maybe *I* might not be so okay.

My fingers didn't have the strength to grab the zipper and pull it open, and I felt woozy the moment I tried to grab it. The adrenaline that had kept me warm while running in wet clothes for the last two hours had finally worn off, and suddenly I realised my own hypothermia had been lurking quietly in the background the whole time.

"Just stop for a minute to regain your strength", it whispered. My legs felt rubbery, and I went to sit down... but quickly stopped myself. Through the foggy numbness, I realised I wouldn't be able to get back up if I sat down - there wasn't anyone left to put me in a sleeping bag after I inevitably fell asleep, so I'd die where I sat.

Leaning against a tree, I bit the top of my jacket collar with my teeth, pressed my palms numbly on either side of the zip's runner, and pulled. The zip opened less than a centimetre, but it was progress! I bit the collar a little lower, palmed the zip runner again, and it slid down a centimetre more. I kept shifting my bite down the collar and palming the zip, and once I could latch onto the opened zip instead of the collar, the rest started to open up quickly. I reached inside the jacket and used the edge of my hand to push the runner down to the bottom, then let out a sigh of relief as the jacket opened and I wriggled out of it. Looking back I only wish someone else had been awake to witness it - I'm sure my jacket-biting muppetry would have looked hilarious if it hadn't been so terrifying.

I wriggled my arms from the jacket, and the movement flushed enough blood into my fingers that I could move them a little more confidently. Rather than struggle with the buttons on my shirt I pulled it over my head, sat down on my sleeping bag to thumb at my boot laces until they loosened and slid my boots off, then wriggled out of my wet pants.

Naked and soaked in the forest, the cold wind snapped me fully awake again as I put my dry gear on, then wriggled awkwardly into my sleeping bag. Within minutes I began shivering - a welcome sign that blood was moving from my chest and head into my limbs again. Then I blacked out.

The next morning I woke with a start, and looked confused when I saw the sun was already up. Sleeping after sunrise was unheard of - we were *always* packed and ready to move at least an hour *before* dawn in case an attack started at daybreak. I looked around to see the rest of the troop cooking, shaving, and generally getting ready for the day, and immediately spotted the last two I'd helped get in their sleeping bags - they'd made it through the night as I'd hoped. Suddenly I glanced down to see my shirt covered in dry, frozen blood. How the hell did *that* happen? But there was no time to figure it out, as the training corporals showed up seconds later and immediately started yelling at all of us. They did bring a little good news though: Joe was alive and recovering in hospital.

We'd *all* made it through the night.

As I said earlier, I've had more than my fair share of close calls. I've been caught alone in the wilderness during flash flooding, had to land an out-of-fuel plane during a solo flight gone wrong, and had to crash land a parachute *backwards* after the wind changed suddenly and the chute malfunctioned. Hell, as a Combat Engineer with the Army I once found a booby trap when the cold steel of the tripwire caught me gently under the nose in the dark. You freeze, back your face up just a fraction, and in the moonlight follow the wire to the left to find a simple tie-off. Without moving or breathing, your eyes follow the wire back to the right and... BINGO - THERE IT IS! You study it for a moment, then get to work disarming it.

There's undoubtedly part of you that's scared in these situations, but it gets pushed to the back of your mind as a cold and clinical voice asks "What do you need to do to survive the next few minutes?". Do what you need to do to get out of this, and solve the problem using the resources you have. And if that doesn't work, then so be it - you are lucky to be here in the first place, and being scared of death in those moments won't help you survive them.

But I know I've never been as scared as I was holding on to that plastic sheet. If I'd tripped in the mud or had been too weak to hold on, *I'd* probably survive the night... but the guy we were carrying probably wouldn't have. Screaming to keep the other two awake was nearly as terrifying: I'd said I'd get everyone safely to bed, so I was going to do that no matter what.

Later that day one of the two guys I'd helped last explained where all the dried blood on my shirt had come from: while I was screaming at them, I must have burst a blood vessel in my sinus because I'd been splattering blood from my nose and mouth as they were changing. They'd noticed, but were too terrified of the ginger banshee that was howling at them to say anything, and I'd been too focused on making sure they got to bed to realise my whole face was dripping with blood.

Through my youth, I'd desperately wanted to prove myself through the military. In my teens and early 20's, I convinced myself that being in the military was what I was "meant" to do. The best use of my skills *had* to be as a soldier, and the Army would *have* to see I was special forces material too! I also convinced myself that I could shoot someone if needed. Even more than that, I was convinced I'd *excel* in a role where the job description included "... to close with and kill the enemy".

I *did* excel in training, with both the Australian Defence Force and in the UK. But after six years, it took four of us nearly dying of hypothermia for me to start realising I'd never be able to shoot someone. There's no way I could pull a trigger knowing someone's friends might have to scramble desperately to save them like we'd had to that night - begging someone to stay alive after something *I* had been ordered to do. For what? National interest? Global security? There was no way I could put others through that, no matter where they were from or what they believed.

Looking back, it's clear something inside me started to die that night: the scared little boy who'd been trying to mask his fear of people by carrying a gun. I wasn't afraid of dying, but I was fast developing a fear that my actions might add more hurt to the world. Yet I can honestly say I didn't *truly* stop and come to terms with the dark path I was on until I had yet *another* near-death experience a few months later.

Some time during that -19°C blizzard on Dartmoor, I was bitten by a deer tick. Having removed hundreds of ticks while I was with the Australian Army, I didn't give it much thought as I safely removed it in a shower back at Lympstone, and promptly forgot about it. But within weeks, I'd gone from being one of the fittest guys in a commando troop to suddenly struggling to walk up a flight of stairs. Then one morning in March 2010 - six weeks after returning from Dartmoor - I was rushed into Lympstone Hospital, vomiting and unable to walk. No one, not even me, expected the blood tests to show I was carrying a particularly virulent strain of Lyme disease.

Call me crazy, but today I'm convinced I was incredibly *lucky* to get Lyme disease when I did. Learning to walk again wasn't great, but ten weeks recovering in the rehabilitation hospital gave me plenty of time to contemplate life, the universe and everything.

Three weeks after being cleared to return to training, I finalised my discharge from the Royal Marines and left the military for good. To this day I'm convinced if that tick *hadn't* bitten me, I'd have ignored the growing misgivings I felt after that night with hypothermia. I'd have finished training, then deployed to Afghanistan alongside many of my friends. Many of them handled the job with distinction, but many came back broken by what they'd had to do, and I know I'd have been one of those forever scarred by the experience.

So no, I'm not scared of dying on Mars - I'm scared of fucking up when other people's lives are on the line. I'm terrified I'll survive a close call, but know afterwards I could have done more to help someone else who didn't. I'm afraid of deliberately hurting someone, and their friends having to carry them on a plastic sheet to save them the way we had to that night.

When I stop to consider everything that happened across six years of blowing things up with the military, my appreciation for the fleeting nature of human existence grows ten-fold. Between literally fighting sharks while spearfishing, jumping out of perfectly good aircraft, and scuba diving deep enough that a single breath from the wrong cylinder would kill me; I've certainly had my share of close-calls doing stupid things as a civilian too.

So maybe it's not just about recognising how lucky the *species* is - it's realising I'll be okay with whatever happens when I die, as long as I go out doing something to make the world a slightly kinder place. There's no doubt about how ridiculously fortunate I am to be alive today, because when I joined the Royal Marines at 24 I honestly never expected to make it to 30. Every day of my 30's has felt like icing on the cake, and I know now that when my luck *does* finally run out, it'll probably be while I'm doing something that helps people.

Given my track record though, it's *also* likely to be something utterly ridiculous. Because apparently, I don't want to die *just* doing something to help humanity - I want to go out doing something that boring people tell future generations about as a "cautionary tale". Icarus shouldn't be a warning; he should be a role-model.

And there are very few things that boring, pearl-clutching parents warn their kids about more than getting tattoos. Most of the guys I trained and served with had more tattoos than you could count, but for many years I sneered at the idea of being permanently marked simply for the sake of it - if I was going to get inked, it had to *mean* something to me.

So instead of having some half-assed idea scratched on my skin at 24, I decided I'd wait to see if I could make it to 30 first. And rather than being some "inspiring" quote misspelled in cursive, it'd be something unique that reflected how lucky I felt to make it through my 20's, and how in its warped way, the military had helped me genuinely appreciate life. Had I finished training with the Royal Marines, I'd have been aiming to earn a "King's Badge". It's a patch awarded to exceptional recruits who show exemplary performance and leadership during basic training, and it acts as a fast-track for future promotion. The design has a laurel wreath with King George V's royal cypher "G.R." inside it, and sits on the left upper arm of the commando's uniform throughout his career.

To acknowledge that I *didn't* earn a King's Badge but felt pretty damn lucky for surviving so long anyway, a few days after my 30th birthday I had the laurel tattooed in the same spot a King's Badge might have sat. Having the "G.R." tattooed inside the wreath to truly replicate a King's Badge was never going to feel right though, so in its place I had the artist tattoo the seven-leaf clover that Fry finds in Season Three of Futurama!

In case you somehow missed that I'm a massive Futurama nerd.

A few days after I had this done, I showed it to a friend who quickly warned me that getting tattoos can be addictive. But I just laughed at them - I'd waited the best part of a decade before getting this tattoo, it carried enormous meaning for me, and it still stung like hell. There was *no* way I'd be rushing back to a tattoo parlour. Right?

Within three months my entire left shoulder had been covered with a cartoon of Carl Sagan riding a velociraptor and wielding a lightsaber on one side, while Rick Sanchez rode a honey badger on the other. Not because tattoos are addictive, but because they're fucking cool - tell your kids.

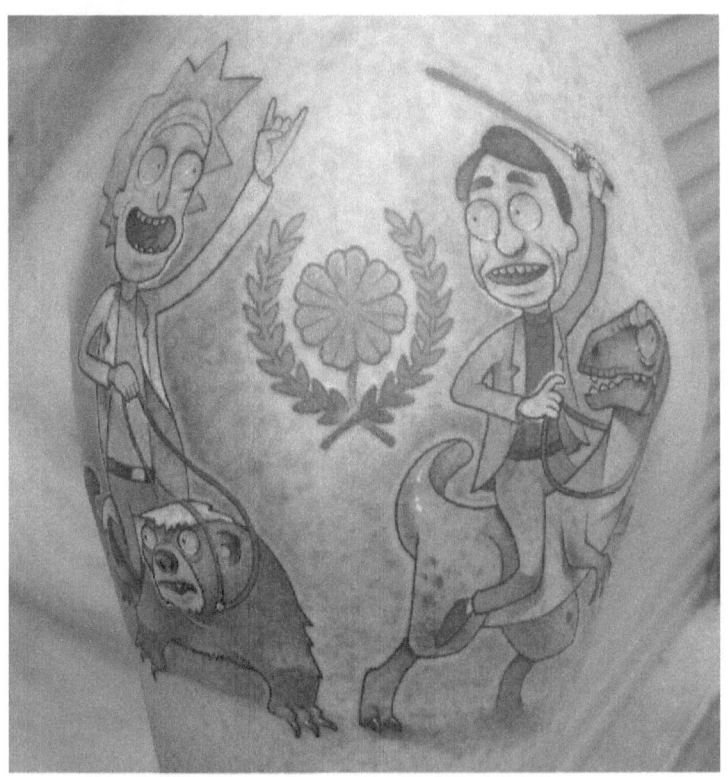

I'm not even remotely sorry

However, it's not just me that feels lucky for making it to 30 - online bookmaker "Sportsbet" seems to think I'm pretty lucky too. When the 100 Mars One candidates were initially shortlisted, they gave me 3:1 odds on for being part of the first crew. Although I'm not sure if I should take it as much of a compliment, considering they also hoped I had "plenty of good knock-knock jokes" because I'll "be up there for a while".

Firstly, anyone trying to make a knock-knock joke on Mars isn't going to get a "Who's there?" back - they're just going to find themselves on the wrong side of an airlock. Secondly, we're going *for good*.

And finally, let's not forget this is the same company offering bets on which Australian politician gets abducted by aliens first (Pauline Hanson pays $5.20) and how the world will end (Ice Age leads the pack at $11).

Call me a skeptic, but I really can't see how a betting company plans to pay out on the apocalypse. I also doubt they'll honour that alien abduction bet either - the moment Pauline Hanson starts screaming about "stopping illegal aliens" the Tralfamadorians are sure to realise their mistake and dump her into the nearest black hole.

Chapter 5 - Attachments

If I *am* going to be one of the first people on Mars, I probably need to spend most of my time and energy preparing to go, rather than blabbering on about how lucky I am while getting cool tattoos. But how the hell *do* you get ready to leave all you've ever known behind? Also, if we're putting four idiots in a tin can and hurling them at a vague red dot amongst the inky blackness of space, precisely what kind of idiot do they need to be?

As soon as I discovered Mars One, I immediately started asking what I'd need to change in my life to have the best chances of dying on Mars. In the short term, this was quite an easy question to answer because I was living in a friend's garage at the time - applying for jobs as a postman while gesturing wildly at the smouldering wreck of my comedy career. Oscar Wilde claimed: "We're all in the gutter, but some of us are looking at the stars". But I doubt Wilde's gutter was on the other side of his mate's rollerdoor when he wrote it.

Standup comedy *was* fun, but it didn't feel emotionally fulfilling and certainly wasn't paying the bills. After a few years, I had to admit I wasn't going to get very far trying to tell dick jokes in comedy clubs like every other cis white dude with an opinion. But maybe I could use my physics degree and life-long interest in science to make *educational* dick jokes at comedy festivals instead! My debut solo show "Apocalypse Meow" premiered at the 2011 Edinburgh Fringe and tackled the science and religion of doomsday. And I obviously wouldn't have discovered Mars One in September 2012 if I hadn't been writing another Edinburgh Fringe show about sending people to Mars one-way.

So the first steps I took preparing to leave the planet were to keep writing the comedy show but to change its tone - shifting from "Hey, this is a thing we should do!" to me standing on a stage saying "Fuck Earth - I'm leaving and you can come too". It would **not** be winning any "critics choice" or "audience favourite" awards.

After consecutive years fighting it out in the maelstrom of the Edinburgh Fringe though, I decided I'd also change tack by writing and performing this new "Mars" show at Australia's National Science Week instead. Both the Edinburgh Fringe and National Science Week run in August each year, so it'd be easy to shift from one to the other. I also knew the Australian science community and media would be far more interested in an Australian potentially going to Mars than folks in the UK would be about some random antipodean trying to escape the planet.

Cracking jokes about Mars also had the benefit of not needing to worry about heat exhaustion whenever I stepped on stage as I had in Edinburgh during the 2012 fringe, where I'd sweated into a giant koala suit while playing the ukulele and screaming at people for an hour each day.

You are *so* welcome

Heading back to Australia unfortunately also meant leaving the garage, which I'd been inhabiting as a kind of ginger troll since the end of the festival had ended. However, I knew I'd *eventually* have to leave the garage if I wanted to live out my dream of being a ginger, cave-dwelling leprechaun on Mars - such are the sacrifices of a wannabe Martian.

Sciency-standup would be an excellent way to engage and excite people about Mars while I was still on Earth, but I suspect a performance art that requires an audience might be challenging once I arrive on the red planet. Having performed to audiences of 2 or 3 people more than a few times in Edinburgh, I know it gets really old, really quickly for everyone. But *writing?* Now that was something I could use as an outlet for my inner weirdo that didn't require human interaction! I wouldn't just write a comedy show about finding the right crew for a mission to Mars - I'd write a *book* about it too!

There was even a template for making the jump from standup to writing. My friend Cameron "Cam" Davies had written *The Human Condition* in mid-2012, which he claimed presented an "in-depth and correct analysis of man's obsessions with sex, violence, money, sex, war, breasts and sex". With a few hundred copies under his arm and a sense of humour drier than the Sahara, Cam turned up to the 2012 Edinburgh Fringe to do readings from his book instead of a traditional standup show. Each day he'd arrive at his venue, sit in a chair and read aloud several of his favourite sections, then sold dozens of copies afterwards. I thought the whole concept was **brilliant**!

While I was still on Earth, I could write science-comedy books about trying to get to Mars and read them at comedy festivals. Then I'd write about the journey to Mars while I was on the seven-month trip there, and once I'd finally arrived in my Martian lair I'd be able to email my "cries-for-help camouflaged as comedy" back to Earth for publication as well! "Meet the Author" events would be awkward with just the same three people showing up, but at least someone on Earth might make a living from editing and publishing my inane martian ramblings. Probably not more than minimum wage though.

At the risk of becoming too meta, it's turned out that writing a book about how I'm preparing to leave Earth has itself been the *perfect* way to prepare. After years of standup, I knew precisely how to write a comedy show about moving to Mars, but the process of writing a *book* felt intimidating and unknown. It'd be so much easier to write another comedy show joking about finding the right crew for the first mission than it would be to tackle the new format, so in 2013 I decided *not* to write the book!

Mars Needs Guitars was written as a comedy show and performed at National Science Week in 2013. And throughout all the event promotion and interviews, I was assuring people a book about going to Mars was in the works. It wasn't. The *idea* for a book was in my head, and I'd written a stage show I was sure could be turned into an entertaining book. But I hadn't taken any tangible steps to turn it into something you could read because I simply didn't know where to begin.

Mars Needs Guitars! toured Australia as well as the Netherlands, where I performed it at a conference hosted by Mars One. "I'll figure out how to publish *Mars Needs Guitars!* as a book when I get back" I lied to myself.

Because of course, I <u>didn't</u> turn *Mars Needs Guitars!* into a book as soon as I finished the 2013 tour and returned to Australia - I immediately started writing *another* science-comedy stage-show about sending humans to Mars for Australia's <u>2014</u> National Science Week!

When it came to *Becoming Martian*, I was determined not to make the same mistake again. **This** time I'd write the book, publish it, and then do readings from it as my "show" at National Science Week - just the way Cameron had with *The Human Condition* two years earlier. Some friends were heading overseas for several months in the middle of 2014 and asked me to house sit for them, so I also had the perfect opportunity to write my masterpiece! There was no need to try and turn *Mars Needs Guitars* itself into a book though, as *Becoming Martian* would combine all I'd learned about sending humans to Mars over the previous two years into one neat package.

With comedy shows it often took years to develop the ideas and themes, but once the concepts solidified I usually needed just two weeks to write the script itself - I expected writing this book would be much the same. By mid-2014, I had all the research and ideas for *Becoming Martian* locked in, and with *months* of isolation ahead there'd be no problem writing and publishing before National Science Week kicked off on August 15th. Easy, right?

In my defence, I *did* write about 50,000 words for the book version of *Becoming Martian* during that 2014 housesit. By July however, it was clear I wasn't even close to having the first draft finished, let alone editing and formatting it into something worthy of publication. The venues for the National Science Week shows were already booked in though, so with the clock ticking I changed gears and turned what I had into *yet another stage show*.

There was no denying I was solidly pissed off at myself. I'd discovered Mars One in September 2012, immediately announced I was going to stop doing comedy shows and write a book instead. But two years later I'd written *another* two stage shows, toured them both globally, and only had half of a shoddy draft for this elusive book. In the end, it'd be nearly **five** years since I'd first spoken about writing and publishing a book about sending humans to Mars before I *eventually* did it. By the time *Becoming Martian* was released for National Science Week 2017, I'd written *three* stage shows about going to Mars and had somehow been selected out of more than 200,000 initial Mars applicants to be one of Mars One's final 100 astronaut candidates.

Why did it take so long? In retrospect it's pretty obvious - I couldn't let go of doing standup. Long before Mars One showed up, I'd been telling people I'd grown sick of it and was looking for a way out. But for years after saying I'd switch to writing books, I was still lurking around comedy rooms because standing on stage and talking about things was familiar and comfortable, while learning to write books was new and hard. Fuck, writing **this** is a nightmare. With every word, you're asking yourself "Is this shit? Am I typing complete shit here? Am *I* shit? Why bother when no one will read it even if you *do* publish. You'll probably delete 90% of this when you edit anyway!". And when I started editing the draft of *Becoming Martian* in early 2017, "deleting 90%" is exactly what happened - I had to write most of the book all over again, as only about 5,000 words of that original 2014 manuscript deserved inclusion.

On the one hand, it'd have been easier to keep doing stage shows about Mars. But at the same time, I knew I *couldn't* keep performing standup at all. Contrary to public appearance, I'm an introvert and find interacting with all of you people and your wet, peach-like eyes exhausting.

I've never experienced the "rush" that other performers talk about - that excitement and energy they get from enthralling a crowd. Speaking to an audience about something incredible or funny can be a *lot* of fun. Still, after performing comedy for more than a decade, I've always left the stage feeling drained by the experience rather than energised. When your job requires you to stand in front of a microphone to say pretty much the same shit over and over again, knowing the only thing genuinely changing is the audience, then even a show about a one-way mission to Mars can become a little tedious.

Put it in a *book* though, and I only ever have to write it *once*. Maybe twice if it's being re-written. But once it's published, I can stop listening to people's face holes flap around with questions and just throw books at them instead. Want to know that thing everyone asks about at *every* talk, even though you covered it in detail in the first few minutes? You'll hear "SHUT UP AND READ THIS" about a split second before my book hits you in the face. Sure, it's easy to have a nervous breakdown trying to get every sentence *just* right. But I've finally started to realise it's not worth the energy, and have decided not to give a shit so much and just do my best instead. If the last eight years has taught me anything, it's to go after what's important and has meaning to you, without fixating on the potential results. It's tough, and you'll often catch yourself giving a lot of shits about things that don't matter. But in the wise words of the Big Lebowski: "Fuck it".

I know I'll sound like every self-help book that's ever plagiarised or misquoted the Dalai Lama, but not being attached to outcomes is genuinely liberating. You stop asking what do I *have* to do to get a good job, to own a house or car, to be respected by my peers, or whatever other bullshit this post-capitalist nightmare has programmed into you.

Instead, you start being honest with yourself and ask what you *want* to do each day, how you can ease the suffering of others, and what you'll give up for it.

But it's tough to let go of all those familiar expectations completely. I knew I needed to write books instead of doing standup, yet it was still a multi-year battle to finally walk away. I'd catch up with an old comedy friend who'd ask if I was writing another show this year, and I'd feel bad saying "no". Or a family friend asking if I had another performance coming up because they'd missed the last one. Or random weirdos spamming me with pleas to bring back Keith the Anger Management Koala **years** after I'd set fire to the suit and posted the footage on Youtube specifically so people would stop asking for it.

Yes, I'm setting fire to a koala suit while wearing another koala suit. Don't think about it too much - I wasn't.

Even three years after I'd said I finished with the performing *Cosmic Nomad* on stage, I rebooted the stage show one more time because someone asked nicely. I *still* didn't want to admit that part of my life was truly over. It's hard to let go when you've invested so much, and yet that's what you need to do to move beyond. As much as you might ask who you need to become to live on Mars, you also need to ask what you're willing to leave behind on Earth.

That's not to say I've given up *comedy* though, because in case you haven't noticed, I'm still *hilarious*. Extraordinarily humble about it too. There's plenty of room in my life for storytelling and writing, but I've had to let go of standup. It was something that helped define who I was for several years, and I'll always be grateful for the artistic framework standup provided as I scrambled through the emotional vacuum I felt after leaving the military. These days though, standing on a stage is not what I need or want anymore, and looking back it feels like vainly hanging onto it after the 2012 Edinburgh Fringe only slowed down my inevitable transition into an author.

Possessions

Much like standup, letting go of my Earthly possessions has had to happen in stages. Although in fairness, it's not like I had much to start with anyway. When I left Meg's garage in Brighton to begin the journey back to Australia, almost everything I owned fitted in a backpack. Now it was a hefty 90L hiking pack, but most of it was taken up by that sweat-encrusted koala suit. All of my clothes fit into a compartment at the bottom of the backpack where a sleeping bag would typically go; while my first aid kit, passport, and wallet sat in the bag's top flap. That was it - my backpack was my house, and half of my house was a filthy koala suit. The only thing I had besides my pack was a $200 pushbike, which for some utterly bizarre reason I paid $300 to have flown back to Australia, and then never rode it again.

My possessions have always needed to serve a purpose. When I left the UK, it felt like I'd simplified what I owned down to the bare bones. Yet things were still a long way from being simple enough for Mars.

Right now, that pushbike isn't doing anything besides rusting in Dad's shed, but it's not like I'll someday strap it to the outside of a rocket to join me on the red planet - it could have been sold years ago to someone who'd love it and use it every day! Hell, that bike should have been given away - it's been of no use to anyone in the last eight years, sitting unloved in a shed when it could have been helping make someone's life a little brighter.

Figuring out little things like this is the beauty of a one-way mission to Mars though - it forces you to ask questions of yourself that you might never ask otherwise. After writing the last paragraph, I googled "Bike Donation" and found half a dozen charities near my parents who clean up old bikes and find grateful riders for them. Some bikes help local homeless folks get around, while others head to developing countries where the push bikes and the shipping container they arrive in all become a sustainable transport centre. After a little digging on Google, I realised that for years there had been a bike drop-off point operating just a few kilometres from where my old bike is still sitting unloved. So why haven't I made an effort to donate it before now?

Part of it is "implied worth" - that bike cost me money after all! It cost me even *more* to get back to Australia too, so at the very least I should sell it and get some of that money back! This is my parents' argument, so for a while I considered getting it cleaned up and advertised. But every time I walked out to the shed and looked at its rusted gears and the cobwebs on the handlebar, I questioned if the time and effort I'd put into selling it would be worth the effort. Easier to just leave it there and not think about, right?

A big part of it is pure nostalgia too, because part of me doesn't *want* to let it go - it's the same bike that got me around Gloucestershire while I worked for an artist during my first year out of the military.

The bike I pumped and sweated on as I rode up and down the hills of Stroud, trying to figure out who I was if I wasn't a soldier anymore. And it's the bike that still carries a sticker from someone I loved a lot stuck on it just before I left the UK. But ultimately that bike isn't coming to Mars with me, I haven't needed it these last few years on Earth, and that relationship ended a long time ago too. Many others would love and cherish a second-hand bike, so it needs to go to them instead.

Strange that this pushbike has lingered on when I've been utterly brutal about clearing out other things that carry far greater personal significance. My revolting koala suit came to its fiery end in late 2015, when I finally acknowledged I no longer needed to pretend to be a violent ex-Army Combat Koala that screamed obscenities at people every day. Most folks who adored "Keith the Anger Management Koala" had no idea that I'd created him as a mask to share things from the military I'd never spoken about and was struggling to process. But after sharing "Keith's" stories in-character during *Keith Looks Back In Anger* at Edinburgh in 2012, I was released from the silencing hold many of those incidents had over me, and I could finally start to acknowledge them and move on.

It's hard to keep playing a conflicted and angry koala character when you're not that conflicted and angry anymore. But you sure as shit get frustrated when everyone keeps asking you to put on a fluffy suit and *pretend* to be full of rage. So I did the only reasonable thing I could think of: I recreated the final scene in Return of the Jedi, where Luke ignites Vader's funeral pyre. Except I set fire to the koala suit while wearing *another* koala suit. Then I stripped out of *that* koala suit and threw it in the fire too.

The whole experience was... *cathartic*.

While I've tried to sell or give away as much as possible, plenty of other things have simply needed hurling into a bonfire. A few months after leaving the military, I started keeping a hand-written journal. Writing by hand slows my thinking down enough to properly process things, while also preventing me from jumping back to "correct" what I've written and tying myself into mental knots in the process. My writing started in bursts and stalls, but within a few months I'd run out of pages in my first journal and had to buy another.

As the years rolled on I kept writing in my journals each day, letting whatever was consuming my thoughts spill out on the page until whatever was quietly lurking in my subconscious started to bubble through. Pages filled, full journals began to stack up, and slowly I realised that keeping a journal had become one of the most useful habits I'd ever developed. Not only does it help me clear my head each day, but reviewing old journals helps reveal consistent patterns and cycles in my thinking that I would never have noticed otherwise!

Regularly journaling and reviewing has been genuinely life-changing, and I'd recommend it to anyone who's after a semi-detached perspective on their life and how they interact with the world. But after you've written those journals and left them to sit for a few months before reviewing, what do you do next? You've picked out all the morsels of wisdom and tried to learn a few lessons through hindsight, but there are still stacks of paper bound in fake leather all sitting in a cupboard somewhere just taking up space.

For a while I'd hoped the journals might eventually wind up in a museum - artifacts for future historians to ponder over while asking "Why the fuck did anyone let this ginger leprechaun in a spaceship, let alone unleash him on another planet?".

Unfortunately, just the *idea* that someone else might eventually read my ramblings immediately started to change the way I was writing. It'd be great to make some witty quip about the "Observer effect" witnessed in quantum entanglement experiments, but this was straight-up performance anxiety. So less like a wave-function collapsing, and more like not being able to pee because some other dude is emptying a firehose in the urinal next to you.

The idea that my journals might have an audience soon started interfering with why I'd started writing them in the first place - processing what was happening in my life for my *own* sake. I tried hiding the full journals away in boxes while telling myself to "stay in the present" and "write for only myself" like some sort of cut-rate Deepak Chopra. But my tolerance for other people's bullshit is only marginally higher than my tolerance for my *own* bullshit, so I knew the best thing would be for the journals to meet the same fiery end as the koala suit.

A year after Keith was doused in petrol and set alight by a ginger idiot in a polyester koala onesie, I went through the stack of 12 journals I'd filled with six years of inane ramblings. I picked out entries that had flickers of what some might mistake for wisdom and photographed them, ready to transcribe and potentially share in the future. Then I invited a bunch of friends around and encouraged them to bring anything that needed to be cleansed by fire - old tax receipts, their ex's broken guitar, unwanted children, etc. We lit an enormous bonfire in my parents' backyard, and I proceeded to throw six years of evidence (sorry, "journals") into the roaring flames.

Some men just want to watch the world burn.

These days I only keep a small cardboard box tucked away in a cupboard at my parents' place, with a few trinkets inside I know *they* would value, along with my two most recent journals. As I finish each journal, I post it back to my parents for them to add to the box, and whenever I visit I open them up to review for snippets of folksy wisdom I'll then photograph and transcribe. Then the reviewed journals find their way to a fire, and the trinket box goes back in the cupboard until the next visit.

Keeping two in the box provides about six months of emotional distance from my writing, so it's easier to review the journals and find patterns in my thinking. However, once they've been reviewed, and the interesting parts photographed, those journals get immediately turned to ash. The journals have shown through the years that my thoughts are always evolving, and they're useful for helping me identify and avoid repeating mistakes. But these days I've no interest in hanging onto them for nostalgia's sake.

Some of the digital transcriptions might be shared one day if somehow I'm convinced my babbling nonsense can help people, but the entries I share will be what *I* am comfortable with people seeing - not merely the raw scribblings I use to clear my thoughts. Future historians might wail at potential artifacts for their "Ginger Space Leprechauns on Mars" exhibit going up in flames, but I'm not journaling for those weirdos. I'm doing this for *me*.

So after years of progressively selling, gifting and burning things, I've now whittled all I own down into a bag and a box. There's the small box of trinkets and journals at my parents' place, but my "house" is my old patrol pack.

At just 25 litres it's a far cry from the 90 litre behemoth I left Brighton with in 2012. But without a crusty koala suit to lug around, it's all I really need.

Inside are a few clothes, a small laptop, a journal, pens, first aid kit, needle and thread to make repairs, my passport, and the item most treasured by hitchhikers and space hobos everywhere: the all-mighty towel. That's it. That's all I've needed to get around the world half a dozen times since signing up to Mars One.

Well, *almost* everything. There is *one* thing that doesn't fit in my pack that I've still carried with me on every single trip: a ukulele.

Ever since I was given a four-string guitar for hobbits at Christmas in 2010, I've loved incompetently mangling treasured songs with its overly bright tone and severely limited harmonic range. I've also spent many evenings creating my own monstrosities to then punish audiences who made the mistake of coming to one of my stage shows. It's not been the *same* ukulele throughout, mind you - I've kept an entire array of different nightmare boxes over the last decade. That first one was a cheap and genuinely teeth-grating soprano given to me by an ex-girlfriend. She was a primary school music teacher who knew I'd had a hideous experience with music in school that had turned me off ever trying to play anything.

Unfortunately, she was *also* an incredibly vindictive human being who was convinced I was too musically incompetent to play anything more complicated than a ukulele or a triangle. She might have been right, but embracing my inner yokel and learning to finger-pick "Dueling Banjos" on a ukulele was the first time I'd ever created music without feeling bad for even trying.

"Not feeling bad" was an unfamiliar emotion to experience around that particular ex-girlfriend though, which is why I was grateful the ukulele turned out to be a "breakup gift", rather than the "Christmas present" I initially thought it was. It's also why I only fooled around with it for a month before giving it away to a three-year-old. He smiled and said thank you, then immediately started using it to club his toy dinosaurs - a moment that brought tears of joy to both us.

That Christmas/breakup ukulele was enough to get me hooked though, so before Ernie started using it for T-Rex batting practise, I'd upgraded to a much more expensive tenor ukulele with a far more pleasing sound. It was a gorgeous, semi-acoustic archtop with mother-of-pearl inlay, and from the moment I picked it up it felt and sounded *right*. My ukulele playing improved immediately, and for the next six months I'd write a new song with each chord I learned. They weren't *good* songs, but at least the neighbourhood cats had stopped howling every time I played.

With my debut solo show *Apocalypse Meow* unfinished and the 2011 Edinburgh fringe only months away, it made sense to direct my new found talent for writing hideous songs into completing a comedy show about the science and religion of doomsday. Less than a week after deciding to include ukulele songs in the show, I was strumming and screeching songs about the Antichrist to the consistent horror of my housemates.

Somewhere in the lead up to *Apocalypse Meow*, I also named the ukulele "Amanda" for no other reason than it just seemed to suit it, and we remained inseparable throughout the festival. As I started writing a comedy show involving a sweating and screaming koala, it only made sense he'd also become a *ukulele-playing* koala that used music to release his not-so-inner rage.

Keith the Anger Management Koala strummed away on *Amanda* at the 2012 Edinburgh Fringe as he howled sing-along classics like "The Explosives Song", "I'm a Psycho", and "Thank Fuck Grandma Is Dead". However, Keith stayed home the night of the fringe I went to "An Evening With Neil Gaiman & Amanda Palmer", which is why Amanda Palmer signed my ukulele "Amanda" rather than having a fuzzy marsupial demon removed by security.

"Amanda" - Tenor Archtop (Kala)

As I started writing shows about Mars One, I also began to question if a full-sized tenor ukulele would be too big to take to Mars.

When Shackleton and his crew had their ship "Endurance" crushed by pack ice near Antarctica, they were forced to abandon ship and camp on the ice. Without any way of calling for rescue though, Shackleton had to make the tough decision to lead those 27 men across the ice to safety. He famously prepared the crew for the long march by ordering them to abandon everything they had except the clothes they were wearing and two pounds of personal items.

However, one of the few exceptions made was Leonard Hussey's 12-pound banjo, which Shackleton referred to as "vital mental medicine". My *Amanda* ukulele may have only weighed a kilo, but if I was going to load into a spacecraft one day, I was determined to try and save weight anyway I could. I'd been investigating "travel" ukuleles that were half the width of a standard tenor, so in 2015 when my then girlfriend asked to borrow *Amanda* for a comedy show she was performing, I saw the perfect opportunity. Instead of lending her *Amanda*, I gifted it instead and bought myself a travel ukulele. And because I'm a romantic moron, I named the new travel ukulele after my then girlfriend and had her sign it as well.

Which is all adorable and sweet until you break up - then you're suddenly stuck with something that's supposed to be fun and cheerful, but you feel miserable every time you pick it up. A few months into the *Cosmic Nomad* tour I used some acetone to clean away where she'd signed it. But the blemish was still there, and that ukulele never felt truly right to play ever again. Much like the relationship, I realised I'd been *putting up* with the sound rather than genuinely enjoying it. *Amanda* may have been larger, but that gave her a far smoother and more beautiful sound, while the slimmer body of the travel ukulele produced such a harsh and twangy tone that I'd soon cut all the songs from the show and even stop playing altogether for a while.

However, the 2016 *Cosmic Nomad* tour eventually took me to the Cincinnati Fringe, where I met an incredible artist who was writing shows to get kids into music. They wanted my travel ukulele to play songs kids would sing along to, and didn't care about its emotional history or acetone smear. There was no way I was giving up the ukulele altogether, but I also knew I didn't want to travel with a full-size tenor again. After selling the travel ukulele to the artist in Cincy I flew to Portland, and within hours of landing I'd bought myself a gorgeous little semi-acoustic soprano.

This one I **deliberately** didn't name though - I wasn't making *that* mistake twice. My "Portland soprano" travelled with me across the US, over to Moscow, before keeping me sane in Northern Israel for several months. Then it was back to the US and through Mexico, on to Ireland and the UK, before eventually returning to Australia with me at the end of 2016.

Once I was back in Australia though, I started to realise that I *did* miss the rich sound of a full-size tenor ukulele. In what was beginning to become a pattern, another girlfriend took an interest in my ukulele, only this time I bought *another* tenor the exact same make and model as *Amanda* for myself, and gave the girlfriend the Portland soprano.

Twelve months later I was still on good terms with the girlfriend I'd given the Portland soprano to, which was quite surprising given my history with this awful little instrument ending numerous relationships. She was still finding the smaller frets challenging to play though, and I was keen to change up the sound of my songs. So we swapped back: she took the larger archtop tenor, while I got back the Portland soprano I'd given her a year before.

I didn't give it a second thought though, because I'd finally decided to stop being such a fucking weirdo who takes ukulele too seriously, and instead decided just to play the damn things for fun. Because that's what they're **supposed** to be - fun.

I **am** still a weirdo about naming ukuleles though. And because that girlfriend's nickname was "Rabbit", I decided I'd finally call the Portland Soprano *Rabbit*, and had her sign it too.

"Rabbit" - Soprano Ovation (Applause)

However, *Rabbit* is far from a perfect ukulele. Just like that girlfriend, I also struggle with it's smaller frets at times - clumsily mangling chords with my awkward troll fingers pawing it petite fretboard. Being a soprano also means *Rabbit* will never sound as rich or full as *Amanda* did. But I've realised it doesn't need to be rich or full - it just needs to be fun.

I'm not playing the ukulele with hopes of becoming some musical maestro, I'm playing because that awkward twang makes me grin like an idiot every time I hear it. There's no denying ukulele can be unbelievably annoying after a while, but that's just because they always sound so unashamedly cheerful. It's like quietly saying "Beep Boop" in an otherwise silent room - after awhile you and everyone else is going to get sick of it, but at least you'll have a stupid grin on your face when someone finally snaps and screams for you to stop disrupting the funeral.

I'm not sure Shackleton would necessarily have taken a ukulele over Leonard Hussey's banjo, but I certainly would. One of the questions I get the most about going to Mars is what I'd take with me on the trip, and the truth is a ukulele is all I'd probably want. Maybe throw in some extra strings just in case one of my crewmates cuts them in a ukulele-induced rage. Photos of family and friends can all be stored digitally, along with favourite books and films, and we'll be able to email back and forth with Earth regularly too. Any diving gear I have won't be much use on a cold and desolate planet, and I suspect digital paper is going to improve enough in the next few years that I won't need to write in or burn paper journals on Mars either.

Before Apollo 11 launched, Neil Armstrong dryly replied to a reporter that he only wished he could take more fuel to the Moon. But since we're going more than 500 times further and then **staying there**, I figure we can probably find somewhere on board to stow the *Rabbit*.

Happy

Years of trimming everything back has radically changed the way I live on Earth too. Not in that tepid inspirational speaker "Oh geez, it really makes you value what you already have!" way that's supposed to make everyone feel warm and fuzzy about "family values", "building community" and other tired agent-of-mediocrity bullshit. No, the freedom I've felt trimming what I need down to a carry-on backpack has only encouraged me to trim down my presence on Earth even further.

Once I started to look at it seriously, I realised a lot of the stuff I was "carrying" wasn't stuff at all - it was just a long series of assumptions. Even though I'd signed up for a one-way mission to Mars, there was still this overlying assumption that my life would follow a familiar path: find a "stable" job to make a "living", meet someone and get married, father a couple of mewling fuck trophies, build a house in the suburbs to shelter them, keep working to put the little oxygen thieves through school so they'd "have the best chance in life", and eventually die hoping there's an afterlife and I'd led a "good life" so all this bullshit was worth it.

Fuck the "good life" or the afterlife - I want to live a radical life now, doing awesome things and dying doing something so stupid that other people will use me as a warning to their kids. *That* is the legacy I'm aspiring to.

Once you cut the assumption that you need more than a backpack of clothes to travel the world, you also start questioning those other underlying assumptions too. Do I need to work a regular 9-5 job if I'm not trying to own a home or feed a family? NOPE. "But... but... what about your retirement, Josh?".

Fuck my retirement - it'll be on Mars, and I'll be more concerned with my air supply than the returns on my superannuation fund. Also who the fuck under 50 still believes working a "regular job" will ever lead to retirement? Even if I weren't leaving the planet, it'd be at least 2050 before I reach even the current Australian pension age of 65. If I'm still on Earth by then, I'll be far more concerned about sharpening sticks to fight off radioactive acid-breathing lemurs, as this peak-capitalist hellscape we call civilisation finally collapses in a pool of its own climate-change-denying vomit.

There's not going to be any "retirement": either we get off this rock and learn to live sustainably in space, or we'll heat this place up enough that the planet will casually bitch-slap our puny species out of existence. These "Save the Planet" and "Save Earth" people have it all wrong - it should be "Stop being assholes or we'll all die". Ice caps will melt, all the gas-spewing humans will drown, Earth will cool down again because there's no one left to fuck things up, and then the Orcas and dolphins will battle it out for global supremacy.

If for some reason all the humans suddenly died - asteroid impact, global zombie pandemic, a plague of mutant species of carnivorous plants, whatever - then most evidence of our existence would be wiped away in a few thousand years. That's barely a flicker in the 4.5 billion year history of this planet. So Earth itself will be just fine without humans, regardless of how much you or Ozymandias have saved in your respective "retirement funds".

I only need to earn a little bit of money to feed myself while I travel around trying to get us off this rock before we all die on it. All those assumptions about "regular jobs" and living "regular lives" are just that - assumptions. I didn't agree to any of it, and it seems obvious now that they never applied to me in the first place.

Still, it's taken years of saying "I'm moving to Mars" and thinking about the implications of that statement to recognise those ingrained life assumptions and then break them.

Beyond where I live, what I do, and if I breed, there are also cultural assumptions that plenty of people take for granted. "Always respect your elders" might be a good general rule for life, but what if your elders are bigoted assholes? Or the assumption you should "Always listen politely when someone is speaking". We teach kids cute little sayings like "You have two ears, but only one mouth" to reinforce the idea, but have you ever wanted to use an industrial belt sander on your genitals rather than keep nodding along to someone droning on about their boring life? Maybe I've just met a bunch of generally shitty people who've used "age" and "experience" to justify their obnoxious behaviour. I'm all for giving people the benefit of the doubt, learning from those who have wisdom to share, and being generally respectful to people no matter who they are. But my time on Earth is limited, so I'm not planning to waste any of it around assholes of any type.

One of the other assumptions is that you should live your life doing things "that will make you happy". Everyone. Just. Wants. To. Be. HAPPY! Happy! Haaaaaaaaaapppppppppppppppyyyyyyy!!!

Here's a fun fact though: Happiness is complete and utter bullshit.

That feeling you had walking down the aisle on your wedding day, or the feeling *I* had eating pork belly glazed with whisky and honey last Friday? That's joy and orgasmic bliss respectively, not happiness, and it's fleeting. What most people describe as being "happy" is really being "content" with life. You've got the house, the spouse, the job, the car - you're all set!

Things couldn't be better, so good for you! You keep doing that thing you're doing, and never change it. Nothing will ever change. Nope, nothing. Everything is always going to be just like this. Forever. Just imagine the years will keep going on and on and on and on...

And nothing will ever truly get better, because nothing ever changes. "Happy" is the false sense of status quo in an ever-changing universe. You'll be relentlessly happy, happy, haaaaaaaaappy... until you have a nervous breakdown or a mid-life crisis as you realise you've done nothing of any real meaning with your life. Maybe it's those first grey hairs, or a bald spot reminding you that things do change, and one day you *will* die. You've chased contentment and security through the guise of "happy" your whole life, and it feels like the apocalypse when something unexpected bumps you from the stable groove you've carved out.

In many ways, I'm lucky I had my nervous breakdown at 22 and realised I'd rather kill myself than continue to pretend I was "happy". More than a decade after I nearly committed suicide while working a quarter-of-a-million-dollars per year mining job, I still occasionally find myself tempted by the dream of a "normal happy life" where I'll be "safe" and "secure" like everyone else.

But those jobs aren't safe or secure. There are no safe harbours, only temporary ones. You can lie to yourself for a while that having this job or relationship or thing is "all you need". But that job, relationship or thing *will* change over time, and so will you. So you can either acknowledge that, embrace the full spectrum of human experience and evolve with it... or you can lie to yourself, say you're "happy" with how things are, and desperately try to ignore the continually changing universe around you.

I haven't wanted to be "happy" for a very long time, but signing up for Mars One helped me acknowledge that I do want to do things every single day that I find meaningful. To do something each day to steer some of that constant change to help make life better for everyone. Contribute to things that interest and excite me, and do them regardless of whether I'm "happy" doing it. Living isn't about finding a way to be safe and content until you die - it's about seeking meaning, embracing the incredible, and burning brightly to help light the path for others. If I've only got a limited amount of time on this planet, I'm going to do all I can to find meaning in my life by enriching the human experience, rather than seeking some flimsy sense of "happiness".

Because rather than trying to simply "be happy", I have an incredible opportunity to question everything about life and find meaning in all I do - to live by principles that I've developed and tested for myself, rather than just blindly accepting the prevailing dogma. I won't bore you with a complete list of my "Things To Live By" here, because a) they're my principles, not yours, and b) it includes things like "Everyone is going to die, but I intend to deserve it". If you're still interested though, I've included it as an appendix - you've been thoroughly warned.

However, one thing that I have always tried to live by is to stay true to my word. From a young age, I realised I could be an amazingly brazen liar if I wanted to be. But also I'm far too lazy to maintain a web of deceit for long - it's just too tiring trying to remember which version of reality you've told different people. Plus, you then have to carry that lie forever, and regularly make up new lies to maintain it. Which then requires even *more* lies to sustain too. Frankly, I've got better shit to do with my time and energy.

Sure, you occasionally have to tell people things they don't like. And they might not take it well and sit down in a paddock, refusing to move until you apologise for telling the truth. But you're not going to apologise, because you're not sorry for being honest, and then you'll leave them in that cow paddock to hopefully be eaten by wolves. I've found keeping my word has come at the cost of friendships with a few folks who valued their pride over integrity. But once you've excluded the manipulative toddlers, it seems like most people would prefer brutal honesty over seductive lies too. Being unflinchingly honest has become something I value in both myself and others, mostly because I get bored trying to sort truth from fiction when someone is trying to mislead me.

Where I think a lot of folks go wrong though is confusing "being honest" with "being an asshole". As someone who is quite adept at both, I've realised the difference is assholes have opinions that they *have* to share. There's the old saying "Opinions are like assholes - everyone's got one", but I've found if you don't force your opinion on people then they're less likely to call you an asshole. I also presume the reverse is true: if you don't force your asshole on people, then they're less likely to have strong opinions about you.

These days I don't often care enough to have any opinion on whatever latest thing everyone's excited/upset about, because I'm focused on leaving all of you behind for another planet. Talk to me about sending people to Mars one-way, or how we should stop doing shitty things to destroy the atmosphere we breathe, or how we can avoid everyone starving before we become a predominantly space-faring species. *Then* I might give you an opinion, but chances are I probably still won't be bothered. The reality is I'm just not interested in trying to win people over anymore. Putting humans on Mars is a thing I care about, I know why it's important to me, and that it'll help a lot of other people in the process.

Explaining is draining - please just fuck off and let me get on with being a grumbly misanthrope as I do my bit in helping humanity become a multi-planetary species. And the only thing more draining than explaining is arguing why I'm right, so I don't usually bother with that either. When I do occasionally wind up in an argument, it's over something stupid like whether the dress is blue, if it's gold, or if you're colour-blind. I'm usually utterly wrong about whatever it was anyway, so I save my energy by shrugging my shoulders and shutting up. There are plenty of people out there with far more interesting opinions and unique viewpoints than you'll get from yet another straight cis white guy with a science and engineering background, and there's absolutely no reason to be threatened by someone who'll provide a different perspective.

When someone asks what I think about something, then I'll happily share. But time and time again I've learned that if I withhold judgement and sit back to listen, I'll learn a whole lot about something I might not have otherwise known. Loud people provide endless insights into their psyche, along with an entire list of their insecurities if you simply listen to them. So if you want to use all you've learned to destroy them later, then you can! I won't, but once again that's just because I'm really lazy about anything that won't directly help with us getting off this rock.

Chapter 6 - Sexy Space

Being known for your integrity might seem great on the surface. But because that same integrity comes from being too lazy to lie, it also means I won't commit to things I'm not immediately interested in doing. Since signing up for Mars One, I've started regularly telling people "No promises, but I'll let you know once I know for sure". Not because I want to lead them on, but because I know I'd drop their thing in a heartbeat if it clashed for even a second with my chance to walk on another planet. Becoming one of humanity's first ambassadors to Mars will always top any opportunity on Earth that doesn't somehow help me leave it.

As you might imagine, telling everyone I know that I can't fully commit to anything except a cold, dead planet 56 million kilometres away does *fantastic* things for my love life. And that's if we even get to a discussion about "commitment", given I'm so embarrassingly bad about making the first move on a date. Pick up booby traps in the dark? Sure! Jump out of an aeroplane from 14,000 feet? Not a problem! Strap into a space capsule sitting above more than a million kilos of explosive fuel, then set it on fire and launch to another planet? In a heartbeat. But lean in for a kiss at the end of an evening when there's clear romantic interest? Noooooooope. Just go for the hug - it's warm, friendly, and emotionally ambiguous.

It's not that I'm scared of having my heart-broken - that's already happened *far* too many times. No, it's because of the overwhelming terror at misinterpreting things and making a platonic connection suddenly feel awkward.

Humans and their social norms confuse and terrify me at the best of times, which is why grocery shopping is one of the most uncomfortable things I'll do during my remaining years on this planet, and why self-serve checkouts are possibly the greatest invention of all time. No longer do I have to make painful small talk with someone behind a register, as they scan through my half-dozen boxes of frozen salt and pepper squid, a four-pack of "Unicorn flavoured" ice creams, and a bag of coffee beans. You still have to interact at the self-checkouts, but that's limited to the supermarket employee who has to linger under the guise of collecting baskets and wishing you a good day to ensure you scanned and paid for *all* of those boxes of frozen salt and pepper squid.

So my existential terror at making the first move on a date isn't a deep-seated fear of connecting with people because I'm planning to eventually leave the planet. It's just that I'm a socially-awkward introvert weirdo who doesn't know the protocol when it comes to trying to cross the threshold from friends to more-than-friends. Is she enjoying this date? Was she asking to take my shopping basket in a sexy way, or was the husky voice because she has a cold? Should I have used the word "protocol" to describe human interaction? Am I just a very convincing cybernetic organism who got a dodgy beta version of my model's emotion software? Is this entire paragraph further evidence of some as-yet-unnamed social disorder that historians will wax lyrically about generations from now?

Once passed that initial "I'm pretty sure she likes me *that* way" threshold, things tend to move amazingly quickly though. But that hesitation is why an embarrassing number of girlfriends have had to *literally* get me into a bed before I've finally thought: "Oh cool - maybe she *is* into me after all!".

For a first date, one ex-girlfriend took me on a romantic steam-train ride, organised dinner at one of her favourite restaurants, then took me to one of her favourite bars where we talked intimately for hours in a corner. Too late for me to get a bus or tram back to my Aunt's place, we got a taxi to my date's apartment where we watched documentaries in her bed until 3 am. But even after all of that, it wasn't until she turned out the light and kissed me that I finally realised "Hey, I think this woman I'm really into MIGHT JUST LIKE ME BACK!".

Honestly, if aliens invaded Earth and the only way to stop the annihilation of humanity was for me to make the first move on a date with their sexy alien hive queen, then we're almost certainly fucked. Because even then, I'm *still* going to need more obvious signals from her than simply asking about my species as she seductively sways her flippers and licks her mandibles.

So even if I *do* meet someone who'll overlook the fact our relationship has a "leaving for another planet" deadline before it even starts, I still need to get drunk enough to finally mumble "Should I kiss you?" or for her just to jump on me. The drinking option is suboptimal of course, because I'll be bordering on passing out before I get *that* forward. But being physically leapt on can always go terribly wrong too. However, being a Mars One candidate has provided a new and slightly terrifying way to overcome my hardwired inability to act on sexual interests.

I speak, of course, of sexy space geeks.

Laugh if you will, but sexy space geeks are 100% a thing, and for the last few years they've kept turning up at my standup shows to both terrify and tempt me.

It's become impossible to dismiss the fact that some people get so unashamedly excited by space exploration that there is *zero* doubt about their intentions for a little red man who just wants to live on a little red planet, and the whole thing can be a bit intimidating for a socially-inept ginger idiot.

Even so, I'm still a bit of a romantic at heart and don't like the idea of trading off the flimsy notion of being an "astronaut candidate" purely to sleep with women that fulfil my adolescent crush on Velma Dinkley. It also seems unfair to those *not* shortlisted for a one-way mission to Mars - regardless of gender - to miss out on the space nerd fun, which is why I'm now going to teach all of you how to get laid with astronomy.

<u>Note to Astronomers:</u> Yes, I know that human space exploration is *very* different from astronomy, but whining about it won't get you laid. So there's your first tip - you're welcome. Also, if astronomers are going to keep appearing on TV to talk about people going to space, then I'm going to keep talking about trying to awkwardly seduce people by pointing at the stars - quid pro quo.

Let's be clear that I'm <u>not</u> teaching you "the game" for *convincing* anyone to sleep with you - this is about trying to reduce awkwardness while adding a little science education to what might otherwise be a shallow and no-nonsense affair. Call me old fashioned, but I feel like if someone is going to Mars-fan their way into sleeping with a future astronaut for the geek bragging rights, then they should at least learn something about the universe in the process.

Now ultimately, getting laid with astronomy is all about *excellent science communication*. Just like regular science communication, you'll need to:
1) Do a little bit of preparation beforehand, and
2) Not to go too deep into details your audience won't appreciate.

Before your date, you'll want to prepare by working out which way East and West are - generally where the Sun rises and sets respectively. Once you know where East and West are on the horizon, it also becomes relatively easy to find the path the Sun has followed through the sky during the day.

If you're organising this date near the equator, then the Sun will pass pretty much directly overhead. If you're south of the equator in somewhere like Australia - looking up to the sky on a romantic river cruise, where the glimmer of the stars is indistinguishable from the eyes of the saltwater crocodiles creeping up on your barge - then the Sun will have tracked in an arc slightly to the North. Likewise, if you're North of the equator - dreamily gazing into the night from some mosquito-infested bayou in Louisiana, where the glimmer of the stars blends the eyes of the alligators creeping up on your airboat - then the Sun will have tracked slightly further South.

This path through the sky taken by the Sun during the day is known as the "ecliptic", and it's roughly the same path most of the planets follow too. Not Pluto of course, because no matter how many times I have to scream it at people *Pluto is not a fucking planet*. But if you look up to the night sky, work out where the ecliptic is and see a bright point near by, then chances are it's not a star - it's likely to be Jupiter or Saturn. If you see a pale pinky-red "star" near the ecliptic, then you've probably spotted Mars. Sexy, desolate Mars.

If you're in the northern hemisphere and feeling brave, you can also look North (e.g. directly to the right of where the Sun set) and try to find Ursa Major - aka "Big She-Bear" to the ancient Greeks, "The Plough" if you're from the UK, or the "Big Dipper" if your culture needs to make everything sound like a theme park ride. Near Ursa Major there's also Ursa Minor, aka "Little Dipper" because Americans think it looks like a ladle, and that everything that's not a theme park ride should either be food or help serve it. This smaller constellation includes one of the brightest stars in the sky, known by many as the "North Star" because it's so close to the North celestial pole.

You can't see the North Star from the southern hemisphere, but you can still show off your shakey understanding of the night sky by finding the Southern Cross! Simply face South (e.g. directly to the left of where the Sun set) and then look for those distinctive four stars that you'll instantly recognise from the neck and back tattoos of Australian white nationalists[1]. In the sky near those four stars, you'll also see two much brighter stars. These are known as "The Pointers" because they "point" to the South celestial pole. If you imagine a line passing perpendicular between the Pointers, then imagine another line running through the length of the Southern Cross, then the point those two lines intersect is the *South* celestial pole!

Both of the "Pointers" are also triple star systems, with the pointer closest to the Southern Cross known as "Hadar" and the other being Alpha Centauri. Alpha Centauri is also home to Earth's nearest star system, with at least two planets.

---

[1] If your date happens to have a southern cross tattoo it's probably better to forget about any attempt at romance through astronomy, and personally I'd be considering an in-flight abort on the entire evening.

Unfortunately, most white nationalists are unaware of all this as they're having the constellations etched into their skin ahead of their next "Reclaim Australia" march. Maybe if they understood Earth's insignificance in the cosmos, they'd stop being such ignorant cockwombles.

If this "lines in the sky" and "face south" business sounds too complicated though, then there's an app for that! Android users can head to the Google Play store and install "Sky Map" - it's free and gives you an up-to-date view of the night sky with all the planets, constellations, and galaxies wherever you are, in exchange for tracking data on wherever you are. For all you savvy iPhone users there's "SkyView" - it has half the features and costs $1.99, but as an Apple user you're probably used to that.

So you've met your sexy space geek, and installed an app to learn the night sky - now it's time to kick it up a notch with a romantic picnic dinner on the beach! You can watch the sunset, then impress them with the astronomy facts you looked up on your phone 10 minutes before they arrived. You know, because you're super prepared for this and definitely *not* a robot that's continuously confused by human emotions.

If you're on a beach with the ocean to the West, then you can look out over the water to where the Sun is setting. Just as it drops below the horizon, look for the green flash - it's a real thing caused by the refraction of the Sun's light through the atmosphere, and sadly *not* someone's soul being sucked into Davy Jones's Locker. As the sky slowly gets darker, the first "star" you're likely to see will be just above the setting Sun. But it's not a star - it's the planet Venus.

Before you get all carried away and start frothing out astronomy facts, *this* is where you need to turn on your talents as a science communicator. The whole art to seduction through astronomy isn't in wowing your date with obscure details - it's in providing just the <u>right amount of interesting information.</u>

Don't get bogged down in the details, and also don't feel like you have to impress them by knowing *everything*. If you're on a stargazing date and you've done a little bit of research into a few different constellations, then you're probably doing okay already! Just chill out and share the cool space stuff you *did* look up on Wikipedia right before she got there, and certainly don't look things up on your phone once the date starts. You don't always have to be right, so if you don't know something, then just say so!

Likewise, if your space geek excitedly shares an astronomy fact that doesn't sound quite right, then for the love of Crustacean Jebus don't try to correct them - this isn't Twitter. There is no need to shoot down a pants-tightening conversation about astronomy with "Well ***actually*** most stars have burned out long ago, but they're so far away we haven't seen them die yet. We're really looking at the *ghosts* of stars, not *actual* stars". Don't be that guy. There is absolutely no need to fuck this up by needing to be "*technically"* correct.

If you hold hands and dreamily share how every element in our bodies began in the hearts of stars billions of years ago, how those stars eventually exploded in magnificent supernovas, and that we are all made of starstuff, then you might get laid. But if your inner over-explainer then kicks in to add that "Yeah, we're all made of starstuff... but then again, so is garbage" then ***you'll probably spoil the moment.***

If you tell a space geek that you're an astronaut candidate for the first human mission to Mars, and leave it at that, then you *might* peak their interest in the pants department. But if you add that it's a *one-way* mission to Mars, that you'll use your own shit as radiation protection on the way there, and that once you arrive on the red planet you'll have to live underground like some kind of mole-person for the rest of your miserable life... then unless they're into some *really* kinky role-play it's probably not going to work out.

I don't actually suggest telling people you're an astronaut candidate in an attempt to get laid though, because it's certainly never worked for me. The last time I went to a networking event a woman walked up to me, asked what I did, and I replied: "I'm an astronaut candidate for the first human mission to Mars". She told me it was the worst pickup line she'd ever heard, then walked away. I legitimately *wasn't* trying to pick her up, but it did make the "How to be an Astronaut" workshop I was running later a tad awkward.

No, I've always had to slowly seduce people with my unique blend of adventurous psychosis and social non-sequiturs. I only mention telling people you're an astronaut candidate to get laid because I know someone *else* who's done it. And they didn't just pretend to be any astronaut candidate - they pretended to be *me* to get laid! Better yet, the girl this wannabe-Josh tried his luck with eventually became his girlfriend. It turns out this guy is a science communicator who looks a bit like me, and when they'd first met at a party she'd immediately blurted out "Oh my God, are you Josh Richards? The guy going to Mars?" before he'd had a chance to tell her his real name. So he just *ran with it* and decided to become Josh Richards for the evening to try and get laid. Kind of a weird story for her to tell me in retrospect, but thankfully she figured out he wasn't the "real" Josh Richards *before* she made the decision to sleep with him that night.

I should be a little disgusted that other people are pretending to be me to get laid at parties, or even that people can get *caught* pretending to be me and *still* get laid. But the truth is I'm a little bit proud of them. To shamelessly say "Yup, I'm Josh Richards" and brazenly go with it, get laid, *and* establish a long-term relationship? I can't commend the deceit, but I'm legitimately impressed by the sheer audacity of it all.

Something Real?

Just because I'm planning to live on another planet doesn't mean I have to cut myself off from connections that go deeper than a scary space nerd fling. None of us will live forever - on Earth, Mars, or anywhere else. But that doesn't stop people from emotionally investing and falling in love.

So while short-lived space geek hookups are both terrifying and fun, I've never stopped looking for more meaningful relationships. My time on this planet might have a deadline, but that hasn't stopped me craving that all-consuming kind of love that makes every sense come alive. The kind that completely overwhelms you, while bringing out the very best in you too. That frankly gross kind of love where you want to be a better version of yourself every day for them, and who's equally inspired to live their most joyful life every day because you're part of it.

One of the most common questions I get about going to Mars is "What would you do if you met the love of your life? Would you leave them behind?". It's an amazingly intrusive question, yet people seem to think they're entitled to an answer because I'm doing something they've never imagined before.

I always find it strange that the same people who say they're "just imagining what it'd be like to leave a soulmate behind" then get all huffy when I ask them: "So if the Magic School Bus crashed in the Andes, which of the kids would you eat first? Ralphie looks like he'd be a meal or three. How about that whiny little bitch Arnold? Or do you start on Ms Frizzle and have the lizard as an appetiser?".

However, the truth is a few years ago I met someone who I legitimately considered staying on Earth to be with. Before we met, I'd deliberately closed myself off to meaningful relationships - they just were too hard to maintain given the weirdness of life as a Mars One candidate, where every day was a constant reminder of the temporary nature of our existence. Yet here was someone so extraordinary and vivacious it was impossible not to fall completely and hopelessly in love with her.

She was hysterically funny, yet it was a positive and uplifting kind of humour that shone a light on the dark bitterness I'd often try to pass off as "comedy". She was hyper-intelligent but never used it to cut people down. Aaaaaaand she was gorgeous - a break from my usual taste in Velma Dinkley look-alikes, but undeniably beautiful.
She seemed to burst with energy and life at every moment. I'd essentially met a unicorn, and was utterly smitten.

That's not to say there weren't early warning signs though. In the wise words of Bojack Horseman "When you're wearing rose-tinted glasses, all the red flags just look like flags". For starters, she was a comedian. Dating a comedian might seem appealing on the surface, but they can also be less than great when they're not performing.

While most humans are descended from apes, many comedians are more closely related to the starfish: pretty to look at, but horrifying to watch as they turn themselves inside-out to consume all that is beautiful around them and turn it into comedy. Have you ever sat on a beach to watch a starfish eat something? No matter how many times you say "Oh, this one will be different" it's still like watching John Carpenter's "The Thing" but in horrifying slow motion and with sand instead of snow.

There was also the minor issue of her living on the other side of Australia. Although in fairness, I have a long history of travelling unreasonable distances for emotionally unavailable women. For the first few months of training with the Royal Marines, the only time we'd get off each week was four hours on a Sunday afternoon. Most guys would stay on base and relax, maybe read a book or catch up on sleep. But *I* would bolt out the front gate at Lympstone a few minutes after midday, get on a bus and ride for an hour into Exeter, then walk for half an hour from the bus station to my then girlfriend's apartment.

I'd see her for about 15 minutes, then reverse the whole damn trip to be back at the training base before the 4 pm curfew. It's not even like it was to have sex: we'd have tea and talk about the week, then I'd kiss her goodbye *and have to leave again*. Part of me wonders why she never organised to meet me at the bus station so I didn't have to lose an hour walking, but the rest of me knows *precisely* why that never happened.

Likewise in 2011, I was smitten with a girl from Louisiana who I'd met at the Edinburgh Fringe a few years before, and we'd been writing to each other in old-school hand-written letters for months. I'd sworn I'd never set foot in Louisiana again after a terrifying trip in 2005 that involved a demonic cat and a family "clan" all holding hands around a cake to sing happy birthday to Jesus on Christmas Day. Yet here I was, organising a round-the-world comedy tour that *just* so happened to stopover in Louisiana. When I did eventually walk out of the Baton Rouge airport, she told me to put my bags in the car without a welcome hug or even a "Hi!". I laughed at the sheer awkwardness of it all, then looked at a huge "Welcome To Louisiana" sign and said to myself: "I've made a huge mistake".

But not *enough* of a mistake to learn from it, because a few years later I suddenly found myself moving across Australian for a *comedian*. An extraordinary, sexy and funny unicorn comedian. But still a comedian. There was one other minor red flag that I was willfully ignoring too: **she already had a boyfriend**. Although if I'm honest, the engineer in me just saw that as a problem that needed a solution! These were *all* just problems to be solved, right? **RIGHT?!**

Although it turned out the boyfriend "problem" had an unexpected solution. She started throwing around big words like "polyamory" and "consensual non-monogamy", but as we've previously established I'm confused by most human interactions, so I didn't know what the hell half of this stuff was supposed to mean. There was one element I could wrap my little walnut around though: *sharing*.

Now *sharing* is not something I'd ever experienced in romantic relationships before, but given how utterly smitten I was it's no surprise I didn't hesitate to give this "poly" business a go. It wasn't conventional, but since when have I ever been fussed about being "conventional"? I've signed up for a one-way mission to Mars after all - "conventional" isn't an adjective I get to keep in the dictionary of my ridiculous life any more.

Starting any new relationship is always exciting, but having always been a deeply loyal partner, I'll admit that dating someone who already had a long-term boyfriend was entirely uncharted territory. But what is love if you're not open to new experiences with someone special... and their boyfriend?

I'll clarify that I never met him - to me he was just someone who was also sleeping with someone I loved. The internet informs this is known as a "vee", where she was the "hinge"... I assume because we were both slamming her door? And if I'd also been into her boyfriend, it would have been a "triad". On top of all this, you've got "quads", "primary" and "secondary" partners, "nesting" partners, "unicorns"... but not the same kind of unicorn, a different type of... whatever, shut up - polyamory has a lot of terminology for something that's just supposed to mean "many loves".

As odd as this whole thing might sound, being in a non-monogamous relationship forces you to ask questions about your connections with people that you might never have asked otherwise. A bit like a one-way mission to Mars, it forces you to work out what you're genuinely okay with and what you're really not. And it turns out some things that made you feel sick to the stomach just imagining turn out to be complete non-issues after all.

It's all a bit like sharing a fetish. Some stuff you can just gently work your way up to, while there's other stuff you know there's no right way to suggest. The first time you sleep with someone, it'll probably be pretty vanilla, but you also might notice a few tell-tale signs of freakdom. The next time around those kinks might emerge slightly more - gentle bum taps might become firmer spanks, maybe playing with hair becomes light tugging. As the relationship continues, things keep progressively building up until one day you realise you're paddling someone with a King James Bible while screaming "THE POWER OF CHRIST COMPELS YOU" and your housemate has to yell out the safe word from another room so you'll keep the noise down. These are what I'd call "Incremental" fetishes, where you gradually push comfort zones until you hit a point where someone involved acknowledges that any further is *too* far.

However, there are also "Leap of Faith" fetishes. There's no obvious way to work up to these kinks - you just have to come out and say you want something weird, then hope your partner doesn't look at you in horror. Like sexy pooping. Why? Why is this a thing? I'm trying hard not to judge - I just don't understand why it's a thing for *anybody*. There's no conversation you can have that politely raises the idea of shitting on someone. None. The one time someone *accidentally* pooped on me during sex, I immediately ran to the shower to wash it off, then we both agreed to never speak about it again for the rest of our lives (sorry).

Emotionally though, some things that initially made your heart hurt and consumed your every waking thought might turn out to be the best things to ever happen. In my case, it was the idea of one-night stands that made me realise I could still be loyal without being possessive of my partner.

It took a while to get my head around it because I'd always been told to "find the one", but here I was going to *Mars* where I'd be living with three other people - the chances of finding "the one" on a shortlist for Mars are pretty much nil. With a crew of just two men and two women, even the possibility of a friendly threesome is pretty slim unless the other dude is okay with sitting in the corner with a video camera.

Yet here I was, completely and utterly smitten with someone who *already* had an amazing life on Earth before I came along. Who already had someone who loved her, and who's love wasn't diminished just because her *attention* was being shared. I didn't own her, so why should I have any say in where she directed her affection? All that mattered was that she loved me, and I soon started to *like* that the woman I loved had someone else who loved her too. She deserved to be loved by *lots* of people. Not in a "gangbang at the old warehouse" kind of way, but in an "emotionally loved by lots of people" kind of way. So an *emotional* gangbang. The cool polyamory kids refer to this as "compersion" - experiencing joy in the knowledge your partner is finding bliss in the company of others.

Once I'd gotten on board the compersion-train though, an incredible thought dawned on me: if I was this cool with someone I loved being in love with someone else, then could *I* open myself up to multiple romantic partners as well? Might it be possible for Josh "Fuck humanity, I'm going to Mars" Richards to have *more than one girlfriend at a time?!* Six months earlier, I'd accepted a future with *zero* girlfriends. But now, provided I was open with everyone and had active consent, nothing was stopping me from loving anyone in whatever way we decided! I could have **<u>ALL THE GIRLFRIENDS!</u>**

Because ladies and gentlemen, it turns out love isn't a finite resource that needs to be scrapped and bickered over - figure out what you're looking for, and if you're into someone then say so. If they're interested, see where things go. If they're not, then that's cool too. Just move on and don't be a creep about it, thinking you'll "win them over eventually". Everyone gets taught in kindergarten to share and not to fight over what others have - I've just found that approach works well with my intimate relationships too!

If you're non-monogamous and meet someone who's mutually interested though, then you need to tell them early on where you stand on exclusivity. If they decide to go with it and then want to make things monogamous later, then that's for the pair of you to discuss. But make sure you're upfront about what kind of relationship you want from the start, and keep asking what's true for *you* as things progress. This is not an excuse to be a selfish fuckwit - it's about understanding who you are deep down and being honest about that with the people you invest in emotionally. If you're honest about what you need in each relationship, but find it doesn't mesh with the needs of someone else, then that's okay too - you were okay before this person came along, and you'll be fine afterwards if things have to end too. Find a way to connect with people individually, value each connection on its own without external interference, and don't let anyone else dictate which romantic interests you're "allowed" to have. Simple!

Fuck, maybe I'm not polyamorous after all - perhaps I'm just a full-blown emotional anarchist.

Ultimately, this is all about loving openly, honestly, and fearlessly. The fact that my polyamorous comedian unicorn also loved someone else didn't detract from what we had, and in its own beautifully weird way it actually made things *better*! One day I'll leave the planet, and even before I leave I'll have to spend an extraordinary amount of time in a mock habitat *pretending* to be on Mars.

Why limit yourself to loving one slightly unhinged ginger leprechaun and missing him when he's away training for his Mars mission, when you can *also* experience love with other people who are a lot closer? At the time I felt incredibly lucky to be loved by someone so incredible, and I loved her even more for helping me realise what I truly value from relationships. At the time I genuinely thought I'd found the Leia to my Solo. The Fenchurch to my Arthur Dent. The Leila to my Fry.

But it wasn't to be. She wasn't Leila, and I'm *not* Fry. Instead, it turns out I'm *Lars* - Fry's doomed clone from an alternate timeline with a shaved head, a goatee, and a tattoo of Bender smoking a cigar on his right ass cheek!

It's like looking in a smelly mirror.

As a side note, people often ask me if this hurt. YES! It felt like someone was carving into my ass cheek with a red-hot scalpel. The other question I get is "Did you get the Bender tattoo when you broke up?", and the answer is no - I had it done at the same time as my clover tattoo, about two months *before* the end of the relationship. Because in case you hadn't noticed, I'm kind of into space. So I didn't get a robot spaceman scratched into my ass because of a breakup - I got it because *Bender is the greatest*.

I'm not going to go into why the relationship ended, other than to say that we couldn't mesh our emotional needs. Comedy is what she lives and breathes, so making people laugh is what gets her out of bed every day. Conversely, I can't stand people and have only ever used comedy as a cheap form of therapy. Being committed to this "dying on another planet" thing was always lingering in the background too, so when it finally reached a boiling point we called the whole thing off. Although I'm pretty confident only a *comedian* would tell an astronaut candidate she "needs more space".

Chapter 7 - Getting Over It

Given enough time, it's often easy to look back on a failed relationship and see where things went wrong to try and learn something from the experience. We're emotional beings though, and when a connection you've invested so much in suddenly ends, it's sure to leave a wound. You can't just jump straight to the end of the grieving process to reminisce and summarise the lessons from it all either - you first have to acknowledge the genuine pain you're feeling before you can even begin thinking about the mistakes.

Personally, the best way to get the whole process started is to become utterly pathetic for a while - just totally let go and revel in your disgusting sadness. Start by laying in bed with the blinds drawn, watching Zooey Deschanel being emotionally unavailable in *500 Days Of Summer* as you eat salted caramel ice cream straight from the tub. Don't stop to wipe away those tears either - just weep straight into it and add some more salt to that salted caramel.

Once Joseph Gordon-Levitt has learned his lesson and the icecream is empty, you can put Adele on the stereo and play "Someone Like You" on repeat for the next 5 hours. Make sure you *feel* that pain by taking miserable, sepia-filter selfies on Instagram with one hand, while having a cry-wank with the other.

(Dramatic re-creation)

Heartbreak *is* tough. You invest so much in another, share such incredible emotions, and you *feel* so much - when something that strong ends, it's like the bottom of your world has dropped out. So it's okay to be a little pathetic - at least for a few days. Once the ice cream tubs start to stack up in the corner though, it might begin to dawn on you this entire Zooey Deschanel thing might have gone a little too far - there's only so many times you can watch *Elf* before the whole room starts to smell of sadness.

You might not want to - you might feel like everything hurts, thinking it would be so much easier to curl up under the covers and just die. But NO! You can't let one breakup define you! You've grieved, but now you're going to stop listening to sad songs on repeat like a fucking emo kid - you're going to seize the day and *get back out there!*

Now is *precisely* the time to leap out of bed, rip open the blinds, wash the cum and tears from your sheets, and install Tinder!

Because you've finished the "wounded puppy" phase, and you're ready to swing *way* too far in the opposite direction and rebound like a motherfucker! One of my friends said: "You know, the best way to get over someone is to get under someone else". So I slept with her, and I'll be honest - it helped!

Things can get pretty chaotic in the rebound phase, and it's easy to make highly questionable decisions. When my life has felt chaotic in the past, I've always found solace in the sciences. And so with this particular rebound, I naturally felt drawn to the most romantic science of all: statistics.

Because while the broken-hearted might initially proclaim "She was one in a million", it also means there's 25 *just like her* in Australia alone! There's more than 7,500 worldwide - **BOOK A FLIGHT AND FIND THE OTHERS!** Hell, you dated a unicorn before, so why not aim to find someone even rarer? Forget about "one in a million" and go for one in a *billion*! Fly to exotic locations, go on sexy adventures and introduce yourself to people by saying "Hey sexy Mamma, are you the five-sigma statistical outlier I've been looking for?". WHAT COULD POSSIBLY GO WRONG?

If all the global travel and Baysian probability seems like a lot of work though, you could always go the other way and try to sleep with 100 people who are "one in a thousand" instead. Maybe don't try this in your small country town of 83, but I won't judge how you make the math add up as long as you use protection.

Because when you're desperately trying to convince everyone you're "just fine" while hiding that still-bleeding broken heart, the best approach may simply be to sleep with half a city in a remarkably short amount of time.

Besides, while some of those *terrible* decisions may haunt you for years to come, you're still going to get a shitload of ridiculously funny stories out of it in the process. With a little time and distance, you may even start telling people about the time your date farted when she came, because it's infinitely more hilarious than if you'd been boring and "taken time to be alone" after a breakup.

It also turns out trying to find love has more in common with searching for extraterrestrial life in the Universe than you might expect. Arthur C Clarke gets quoted for saying: "Two possibilities exist - either we are alone in the Universe, or we are not. Both are equally terrifying". But if we *are* alone, then there's currently no way to prove it definitively. Even if you were to check every planet in the Universe and find nothing, aliens might have evolved on a planet after you checked it - you'd have to check *everywhere* in the Universe simultaneously, forever. So much like love, you can only ever say you haven't found ET *yet*, and to keep looking until you do.

There are certainly ways to narrow the search for ET though. Back in the early '60s, an astronomer by the name of Frank Drake desperately wanted to talk to aliens. But rather than connecting with aliens by doing a bunch of acid like everyone else, Frank decided to nerd out and develop a math equation instead.

$$N = R^* \cdot f_p \cdot n_e \cdot f_l \cdot f_i \cdot f_c \cdot L$$

What became known as the "Drake Equation" is a sequential series that uses the probability of each step required for intelligent life to start broadcasting radio signals, and these probabilities are factored together to try and estimate "N" - the number of alien civilisations we could potentially communicate with inside the Milky Way.

$R*$ is the rate that new stars form in our galaxy, which Frank initially estimated at one new star per year. However, current data from NASA and ESA estimates star formation could be three times faster than that.

fp is the fraction of stars that have planets orbiting them. Frank estimated that this was between 0.2 and 0.5, so only one in every two to five stars had an orbiting world. Today though, we know this number is pretty close to 1, and it's pretty incredible to realise almost *all* of the stars you see in the sky have at least one planet orbiting them!

Not all of those planets could support life though, so *ne* reflects the number of planets the correct distance away from their star to be *potentially* habitable. Frank figured that for every star system that had planets, between one and five of those planets would be in the "Goldilocks zone". In 2013, data from the Kepler space telescope suggested it was even better than that - each star with planets probably had at least to three to five habitable ones!

Unfortunately, calculating the rest of the Drake equation relies on a lot of shaky data and serious guesswork. So far the only planet we've got an example of life on is Earth, and we're not even sure if life on Earth started here in the first place! Life might have come to Earth via an asteroid hitting Mars, hitching a ride on an interstellar comet, or maybe Kurt Vonnegut's time-travelling ghost went back 3.5 billion years to fire eight-hundred pounds of freeze-dried jizzum at Earth from the non-planet Pluto. Who knows? We're not even entirely sure what conditions lead to the development of the most basic forms of life, so it makes it pretty tough to work out fl - the fraction of habitable planets that go on to support basic life.

Beyond that, it's entirely guesswork determining the fraction of planets that could support intelligent life (fi), or the probability that intelligent life then develops technology to broadcast signals we might detect (fc). The final variable is L - how long an alien civilisation sticks around broadcasting signals before it stops. Maybe things go quiet because they move to laser communications. Although if they're anything like us, it's far more likely they'll fall silent because they've blown themselves up with nukes, caught a species-ending cold, or committed themselves to a system of economics that urges unchecked growth until their declining planet's environment wipes them out. Who knows, right? I'm just spit-balling here.

Why am I babbling about trying to find ET in the first place though? Well, for starters aliens are awesome, and every single night I look to the sky hoping they'll rescue/abduct me from this planet so I can live light-years from the rest of you. But what's probably more relevant to the overarching structure of this book is that Drake's Equation matches up almost perfectly with Tinder! So instead of trying to calculate N to figure out how many different ET's we can talk to, we're going to use Drake's equation to find friends with benefits who you can probe back.

We'll make R^* the number of people in your swiping radius, which is easy enough to calculate from census data. A broader search radius will cover more ground, but it also means you need to swipe through a lot more data points. Find an R^* value that reflects how far you're honestly willing to travel to make an in-person connection - no point matching with a xenomorph only to realise later they're on the wrong side of the Yarra River.

Next up, we have the fraction of those who are an appropriate age for you to date, aka *fp* or "The Perv Fraction". Now because we're not massive creeps, we'll be operating on the well-established "half your age plus 7" rule.

For example, as a 34-year-old I'd first half my age (17), then add 7 to find my lower age limit of 24. Which honestly *still* seems far too young - they're barely old enough to have had a quarter-life crisis, let alone deal with all that other weird shit that happens just before you turn 30. Likewise, my theoretical upper age is found by taking seven away from my age (27) and then doubling it to find a cougar limit of 54. However, the upper limit is more of a rough guideline, and it's a guideline I believe can and *should* be ignored, especially if Megan Mullally or Sigourney Weaver are involved.

Next, we have *ne* - the fraction of candidates with a gender-identity you're sexually attracted. Sure, the Kepler space telescope found planets around almost every star in the sky. But on a personal level, this can be a challenging variable for people to figure out, especially if they're concerned about what other members of the astrophysics community might think.

You might *claim* to only be looking for alien radio signals from around main sequence G-type stars, but if you think there might be someone home near that brilliant Blue Giant, then why not find out? There are vast communities of supportive researchers out there who *only* look for life around Blue Giants, who'd love to share data with you. Maybe that Cepheid-variable you're looking at tonight is in just the right phase to support an intelligent alien civilisation. It won't be tomorrow though, so unless you point your dish at it *now*, you might never know.

That *ne* fraction can change too, both long term and from day-to-day. Some might have a *ne* of zero, while others are flat out searching the entire sky with a value of 1. In this crazy mixed-up galaxy, only you can figure out the *ne* value that feels right for you, and it doesn't matter what anyone else thinks about your research. We're all just stars waiting for the inevitable heat death of the Universe anyway.

Also, just because *you're* looking for life around particular star types doesn't mean they all want to hear from *your* specific solar system. Hence, *fl* becomes the fraction attracted to whatever sexual spaceship flag you're flying. Growing up, I had to come to terms with the fact that I wouldn't ever be a Red Supergiant. Probably not even a Red Giant, and my refusal to visit stellar nightclubs can often mean I'm not an immediate candidate to check for habitable planets at all.

But I hope researchers who *do* investigate my Red sub-giant system collect plenty of data, and that it's mostly good. The consensus so far seems to have been passable, but maybe they were all just being polite. I guess you just have to keep trying to have fun and do your best until someone tells you to stop trying to make a clumsy astrophysics metaphor about gender and sexuality work.

Of course, just because you've met someone with compatible sexuality doesn't inevitably mean you'll immediately start swapping saliva - you still have to have *fi*, or the fraction with *genuine interest*. No matter what Billy Crystal and Meg Ryan were pushing in the late '80s, I can honestly say I have *zero* interest in trying to sleep with many of my extraordinarily attractive female friends. They're funny, smart, gorgeous humans and we get along fantastically... but that *thing* isn't there. They're just amazing people who I don't feel romantically towards or want to have sex with, and that shouldn't be a bizarre concept to comprehend.

But I don't mean in a "Not wanting to ruin the friendship" kind of way, because I have *absolutely* ruined friendships by telling women I was interested in them. Although in retrospect, any connection that ended because I was honest with someone wasn't particularly strong anyway. I've also told some great friends I had a crush on them, they said they didn't feel the same way, and we *continued as friends like before*. No need to make things weird by continually bringing it up. Nor is there any need to end a friendship because your pride took a hit when you didn't get the answer you hoped for - just keep appreciating your friend for being the awesome human you already knew they were. It's honestly not that hard.

Oh, and before you think you're fooling anyone, please don't have some secret plan to stay friends until they're vulnerable and then hit on them again. Because that's not being "friends" - that's you being a "predatory fuckwit". It's quite different.

You might also find you've got excellent compatibility and a keen interest in each other, but there's a troublesome *fc* or *cultural factor* in the way. They might come from a family with strong conservative values, while you're an economic and emotional anarchist. Or they're Amish, and you can't grow a beard. In my case, I have a history of trying to seduce amazingly smart and sassy curly-haired ginger Jewish women, who then tell me they only date within their community, so I'm left asking "Why not this goy?". There can be countless different reasons why culture and personal priorities don't quite line up to make a connection work, even something as bizarre as her wanting to "grow old together" when you're dead set about dying on another planet.

Fc can also cover the "Sliding Doors" effect - where a quirk of circumstance gets in the way of making a connection, yet somehow it all leads inevitably to Gwyneth Paltrow steaming her vagina. For example, in 2014, I returned to the Netherlands to attend the International Redhead Festival for a second time. And yes, there is a lot to unpack from that sentence. During the Saturday night pub crawl - where literal packs of gingers roam the streets of Breda like wild orange animals - I met not one but *two* charming Irish ladies, who *both* happened to take quite an interest in me. Things were going quite well by the third pub, and I thought my future held a ginger threesome aka "A three-alarm fire". But after I took a quick bathroom break, I returned to the bar to realise they'd both suddenly disappeared.

Looking around confused, I turned to my French brunette ex-girlfriend, who I'd met when she was a festival volunteer two years earlier, and who was now leading our pub crawl group. Asking where the cute Irish girls had gone, my ex just smiled and explained that she'd seen they were taking an interest in me, so while I'd been in the toilet she'd walked over and politely told them I'd be coming home with *her* that night. Which I suppose was fair, considering I *was* staying at her apartment during the festival. But five years later, I'm *still* wondering what could have been, while trying to break the curse around Irish women I've carried ever since.

Finally, our Drake Dating equation would be incomplete without *L*, which defines if your potential match is alive or dead. This is a fairly critical aspect in dating too, and a hard limit unless you're trying to pick up a sexy ghost. If so, excellent communication is supposed to be the foundation of every healthy relationship - reach out through whatever mode of communication works best for you both, be that tarot cards, ouija board or channeling a spirit through Aunt Sally.

Fighting Fermi

When Drake first plugged values into his equations in 1961, he calculated there could be up to 50 million civilisations *just* in this galaxy. Think about that: with more than 100 billion planets in the Milky Way, 1 in every 2000 might have intelligent life we *could talk to!* And that doesn't even factor in that there are 100 billion other *galaxies* outside the Milky Way - the interspecies space phone *should* be ringing off the hook!

Of course, anyone who realises anything incredible and makes the mistake of sharing it gets immediately inundated by a brigade of sad-sack Debbie Downers, which is why as a species we can't have nice things. In Frank Drake's case, he had to deal with a Nobel-prize winning Debbie by the name of Enrico Fermi. Fermi had done some cool stuff with radioactive bricks to create the world's first nuclear reactor and earn the title of "Father of the Atomic Age", but he still felt the need to piss on Frank Drake's ET parade by asking "So where is everybody?". If there are potentially 50 million alien civilisations running around the galaxy, why hasn't at least *one* shown up to say hi? Now there *are* plenty of folks who think ET has been visiting Earth for thousands of years, but for a moment let's ignore that uncle who tries to corner you every Christmas to talk about "The Incident".

The reality is it's incredibly hard for humans with all our self-importance to fathom just how vast the distances are in this galaxy, without even beginning to consider the far larger universe. Space is really, *really* big. I'd quote Douglas Adams on just how big it is, but indulge me as I draw from another modern space poet - Phillip J. Fry: "Space. It seems to go on and on forever. Then you get to the end, and a monkey starts throwing barrels at you".

Even with millions of potential civilisations to communicate with, we've only been broadcasting radio of any reasonable strength for the last 100 years or so. Which means even the first, most embarrassingly weak radio broadcasts our species made have barely travelled 100 light-years into a galaxy that is 200 *million* light-years across.

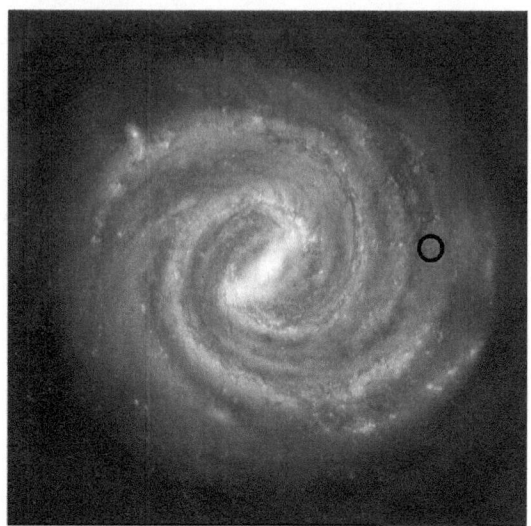

Top-down view of the Milky Way Galaxy

See that small black circle on the right? At the very centre there's a single, tiny dot. That dot is *not* Earth, or even this solar system - that dot is how far our radio broadcasts from the last 100 years have reached across a *13 and a half billion years old galaxy.*

If you represented the Milky Way with a delicious pie, the impact humans have had so far isn't even a sliver of that pie. It's not even a crumb. It's more like someone eating a tiny bit of the pie, digesting it, then farting it in the galaxy's face. That's right; I'll say what no one else will - all the radio transmissions our species have ever made throughout history add up to the equivalent of a galactic cake fart.

Now I'm not sure how much Enrico Fermi liked cake farts, but I do know that bitching because ET hasn't landed on the White House lawn is missing the point. It's no better than someone floating alone in a lifeboat in the middle of the Pacific, crying they haven't found their soulmate on Tinder when they've set their search radius to 1km.

To truly make an ET booty call, what we need is the equivalent of a distress flare or an EPIRB. Which is why in 1974, Frank Drake got together with Carl Sagan and some other sexy space nerds to broadcast our civilisation's Tinder profile into space. Unfortunately, it wound up looking a lot like an intergalactic dick pic:

Worse still, the intention of the "Arecibo message" was to demonstrate "humanity's technological achievement" rather than to start a genuine dialogue with aliens. Which means humanity's first targeted broadcast to ET *was* in fact a dick pic. And it's not like we'd get a response anytime soon, even if aliens *did* accidentally stumble across it: they aimed the message at the M13 globular cluster 25,000 light-years away. So it'll take 25,000 years to get there, and at least 50,000 years for us to get a response. For context, humans were still hunting woolly mammoths and breeding with Neanderthals 50,000 years ago.

So while you can't whine about not finding your soulmate when you've barely made an effort to look, you also can't start flicking dick pics into space and cosmically power-swiping your way through the galaxy expecting to get a match either. Finding ET is about being patient and asking insightful questions about ourselves. Because while half the fun of Tinder is in the search, where it gets exciting is when you figure out what you're looking for and start getting *choosy*.

Likewise, I swung post-breakup from being open to pretty much anything to getting quite picky. It's not that I closed myself off to people the way I had before, it's just that I had a clear picture of what was important to me in a relationship and I started focusing on that instead. Which is a polite way of saying I started looking for unicorns instead of continuing to put party-hats on horses. The best way to spot a unicorn is to draw a picture of what you'd expected them to look like, so I decided to figure out what characteristics I might look for in an "ideal" partner.

Firstly, they'd love whatever they were doing rather than focusing on the final result. It wouldn't matter if it were arranging flowers or flying helicopters, they'd be pushing to do their *own* best rather than worrying about how they compared to anyone else. They'd be self-aware and self-reflective, knowing the only person they needed to impress was themself. I'd want to be around someone who was generally passionate and excited about life. While shit stuff is inevitable, they'd still be grateful to be here at all. Not in that overly tedious "gratitude porn" way - where everyone wanks about how lucky and privileged they are - but in a "I guess I'll just try again tomorrow" kind of way. Grateful that you get better at the things you practice each day, and that you woke up this morning at all to try and do them.

I know I'm instantly cutting my dating pool in half, but I'm also boringly straight too. And I don't mean "I'm telling everyone I'm straight to conceal a dark, worrying suspicion I might not be". I mean I genuinely don't understand what straight women find attractive in men at all - honestly most of us are truly awful, so *what* is the appeal? I know I've disappointed a few of my queer male friends through the years, but after kissing a few beards there's now no doubt I'm a zero on the Kinsey scale, so my ideal partner is going to be a lot more Jessica Rabbit than Roger.

They'd have to be smart too. Like *really* smart. A sexy rocket scientist, a seductive brain surgeon, or maybe both! I've never really understood why so many men are intimidated by intelligent women - I'm already self-sufficient and more than capable of looking after myself, so why *wouldn't* I want to date someone way smarter than me? Someone who might help me understand something I'm not getting, be amazing at things she cares about, and kick ass at every pub trivia night? They're not a threat - they're incredible! There's also substantial research correlating higher intelligence with increased sex drive too, so if other dudes are intimidated by smart women that is *fine* by me. Velma Dinkley essentially kick-started my puberty, so please feel free to pass my number on to your "intimidating" nerdy, curvy and dirty scientist-engineer-librarian lady friends. Also, I'd like to request an introduction to any women you know who build or fly spaceships too, thanks.

I'll admit that even Velma isn't perfect though, but only because she's an ardent rule follower. My barely contained anarchist trait continually seeks others who are secretly harbouring an inner gremlin, ideally one fed after midnight. Someone who blends in and follows the rules, but only until they spot the perfect opportunity to unleash chaos.

The kind of woman you see lurking nearby when you're at a work party, and you're trapped in a boring conversation with Frank from accounting. You make eye contact with her as Frank gushes about his passion for putting tiny ships in bottles, and she smiles at you then quickly disappears. A few minutes later she suddenly reappears over Frank's shoulder to say she's just heard Patty in HR talking about Frank's incredible model of the *Mary Celeste*! As he excitedly scuttles away to find Patty and spark some inter-department romance, your lady anarchist grins again and says "I just turned his stupid bottle ship into a molotov cocktail, and threw it into the HR department". As you look at her with a confused mixture of horror and admiration, she reaches out to hold your hand, looks deeply into your eyes and whispers "Don't worry - I wiped the bottle down and put Patty's fingerprints on it".

I'm not saying I want to date a pyromaniac, just someone who loves throwing fun-filled chaos into the lives of boring people. Someone who wants to test to see if we *really* need bottle ships or HR departments, and knows there's nothing like the rush of excitement you get from disrupting office social events.

Initially, I figured I'd set the bar astronomically high by limiting my unicorn search to women shortlisted for a one-way to Mars too. But I immediately realised I'd be putting unnecessary pressure on an incredibly restricted group. Yes, there are undoubtedly extraordinary women in the astronaut candidate pool, but no one is going to Mars for at least a decade - there's being choosy, and then there's unreasonable. Why limit your romantic options to *just* people crazy enough to sign up for a one-way mission to Mars, when you'll all be on Earth for quite a while yet? There are plenty of other folks who are the right kind of crazy to date a future Martian while you're still here!

Not that it stops people asking amazingly intrusive questions about our future sex lives on Mars. Just like the "Would you leave the love of your life" question, people seem obsessed with the crew having sex on Mars, and I'm always a little blown away by the incredible assumptions they make. Why, for example, would you say "I guess they'll ban you from having sex on Mars"? And why would you be surprised when "Try and stop us!" is my response? What can *anyone* do when we're 56 million kilometres away? Call the cops?

I've worked in plenty of isolated and high-stress environments with mixed-gender groups for several months at a time, and I can assure you there is a mountain of good-natured bed-hopping going on before the end of the first week. If not, you're probably working with buttoned-down psychopaths who will spiral viciously out of control the moment someone gets drunk. People can keep it together and hide their psychosis for a few weeks, but if you're going to be living under each other's feet for months, then it's so much easier if some of you are sharing bunks. Things are going to be a little more close-knit with just 4 of us living together in a tiny habitat on a frozen planet though. With that in mind, it'll be essential to remember that old Martian saying: "If the hab is rockin', maybe try a different angle because you're about to cause a hull breach.

Over time though, I've slowly realised this whole exercise has just been my way of running from an ugly breakup. It's one thing to joke about making things up with a numbers game, or to tell yourself you're looking for someone unique and "special". But the reality is I had a reasonably life-changing relationship end, and there is no quick way to process that hurt. It took a good 6-12 months of running to *other* relationships before I figured out I was still carrying that pain with me and was inflicting it on others in the process. Some truly incredible people who just loved me for me, and couldn't understand why I didn't seem to like myself that much.

Of course, the next tactic was a tried and tested one - turn it all into a comedy show and laugh about it. In 2011 I'd written my first solo show in the aftermath of a horrendous relationship ending soon after I left the military. So naturally, *Apocalypse Meow* was all about Doomsday and finding new beginnings after everything has imploded. Supremely dark stuff, but in its weird, twisted way it helped me come to terms with the life I'd had to leave behind. I'd learned to use comedy as a way to process emotions and written a few more solo shows, but by 2015 I'd become pretty burned out by standup.

But as Henry Miller said: "The best way to get over a woman is to turn her into literature". So in 2016, I wrote and toured *Cosmic Nomad* around the world as my last comedy show. Be under no illusions, it was *her* show. And in many ways, this is her book too.

Looking back, I realise just how much I was still hurting as I toured the stage show. After doing all I could to see the bright side and turn it all into something we could laugh and learn from, I got to the end of a ten-month global comedy tour and had to admit the laughs had all carried a lot of pain with them, and it still wasn't clear what I'd learned from all of it. Those feelings had nothing to do with my ex-girlfriend anymore though - this was all me, trying to figure out why I'd felt so strongly about one person. Why I still felt such a profound sense of loss, when consciously I knew there were so many others who thought I was awesome and legitimately loved me.

It took a lot longer than I'd ever hoped or expected, but I've finally figured out why that particular wound took so long to heal. In the overwhelming rush of emotions I felt during that relationship, I forgot the most profound thing that signing up to Mars One has ever taught me: *this is bigger than you.*

Folks ask "What will you do if you meet the love of your life before you go to Mars?" thinking it's a choice, but their question completely misses the point of Mars One's mission. Because if I gave up on the chance to help humanity become a dual-planet species "for love", it wouldn't be "sweet" or "romantic" - it'd be selfish.

Searching for and finding love is undoubtedly incredible, and meeting an anarchist-pyromaniac-Velma-Dinkley-look alike who build spaceships or does some other equally cool thing would be an extraordinary addition to *my* "universe". But the *actual* Universe doesn't give a shit if I find love or not, because this is so much bigger than me. It's bigger than one ginger idiot's relationships. It's bigger than my friendships, my family, or my country too. Having humans live on Mars will impact the future of our entire species, and I'm just one weirdo cheerleader trying to make it happen.

Even if I did meet someone extraordinary like that, they're not going to make me a more "complete" person. There are very few phrases I detest more than people referring to a partner as their "other half". Don't be a fucking half - be a whole goddamn person. A partner doesn't "complete" you, because that's not a relationship - it's codependency. Antoine de Saint-Exupery nailed it when he said: "Love does not consist in gazing at each other, but in looking outward together in the same direction".

By all means, draw lessons from past relationships so you can develop better ones in the future. But don't fixate on the past. I had to stop and recognise that I'm a pretty fucking awesome person - my *own* magical space unicorn if you will - and that I'm following my own weird and wonderful path, wherever that takes me.

When you're doing something that brings out your absolute best, that makes you leap out of bed and embrace every single moment because you know your time on Earth is limited, then it's hard to comprehend why others would think you need someone to "complete" you. Others are welcome to join the journey, and I'm wildly accommodating when it comes to sharing the path. But if push comes to shove, I'm more than okay with my own company and have never hesitated to go on alone.

More than a few folks I've said this to have told me it's "sad", and that they "hope I eventually meet someone special". But I can only shake my head in amusement, thinking about how lucky I've been to meet so many people I still consider extraordinarily special. Since I signed up to Mars One I've met so many incredible people, had so many extraordinary relationships, and learned so much that it feels like I've compressed half a dozen "love of my life" experiences into as many years.

Plenty of folks will still dismiss that, but I genuinely believe it - I've felt more deeply in love with the partners I've had since I signed up for a one-way mission to Mars than I ever had before, because the keen awareness that my time on Earth is limited has made *all* my feelings stronger. Every time I've met someone I've told them what I've committed to, and *then* if we chose to leap in and have a relationship it's grounded in the knowledge that I'd leave them for Mars tomorrow if given the opportunity. Every relationship is an active and informed choice to share the limited time we have on this planet - acting without expectation, and embracing whole-heartedly what we have here and now.

Although maybe I'm being chill about all of this relationship stuff because in 2012 a fortune-telling gypsy told me I'd meet a unicorn soulmate somewhere between my 32nd and 35th birthday. The prophecy was quite clear apparently: she'd be an Australian archaeologist with curly ginger hair, and we'd meet in Australia.

However, this was especially troubling news at the time because:
1. I had zero intention of ever going back to Australia,
2. I had a well-established history of dating women who *weren't* Australian, and
3. I was somehow back in Louisiana again.

But six months later, I had moved back to my parents' place in Perth. Coincidence? Hardly - I found Mars One and came back to Oz to visit schools and talk to kids about moving to another planet.

If this "Rachel Weisz in The Mummy crossed with Ensign Tilly from Star trek Discovery" space unicorn with an Australian accent is ever going to turn up though, maybe someone should send her a gentle reminder that I'm turning 35 this year - the clock is ticking.

Chapter 8 - Little Martians

Asking about relationships is just foreplay for what everyone *really* wants to ask about though: Will you have kids on Mars? And the short answer is *absolutely not*.

It's partly because any kids we have will probably wind up with jelly bones. There's every reason to believe that growing up in 38% of Earth's gravity will mean Martian kids will have significantly reduced bone density, lower muscle mass, potentially much weaker hearts, and developmental issues with their eyes and spine. But more importantly, why the fuck would I want to ruin a perfectly good planet by having kids on it?

Elton John hinted at it when he said Mars is no place to raise a kid, but the reality is we'll be living in something as complicated and dangerous as a nuclear submarine *on another planet*. Personally, I can't think of many things worse than having a toddler run around smearing shit on the habitat walls and whining about wanting to play outside. It'd be great if we could let them play in the airlock so if the screaming gets too much, I can always turn the volume down by venting a smidge of atmosphere.

Unfortunately, the folks responsible for protecting the Martian environment won't let me vent any of the air out of Little Stevie's Playhouse. Humans are generally disgusting at the best of times, but there are few things more riddled with disease than a runny-nosed 3-year-old. If we're ever going to find evidence of indigenous life on Mars, we'll have to be painstakingly careful about avoiding contamination with bugs we've brought with us from Earth.

Given you only need to walk down the same street as a Kindergarten to be infected with some Ebola-Smallpox-Spanish Flu viral hybrid, the question shouldn't be "Does little Stevie play in the greenhouse, the kitchen, or the airlock?". What we should be asking is "Do we want to ruin any chance of discovering extraterrestrial microbial life on Mars, all because folks on Earth thought it'd be cute for the crew to spawn a wailing, drooling, pumpkin-soup shitting Martian fuck trophy?".

The "babies on Mars" question comes up so frequently though that it's forced me to ask if I'd regret going to Mars instead of staying here to have a family. So many of my friends who've had kids have assured me that "Becoming a Dad changes you", and while they weren't ready when it first happened, they wouldn't ever change it because having kids was "the best thing" they'd ever done.

Which *is* all quite sweet, but it also sounds a lot like Stockholm Syndrome. So many people tell me how great being a parent is, how kids are so full of "potential", and that they're little miracles made out of stardust. But no one talks about what happens if you get a shit one. What if you're the best parent in the world, send your kid to excellent schools and provide the most supportive and loving environment possible... and they *still* turn out to be a dud? After all, I went to high school with mountains of "potential", and the highlights so far have been them producing even *more* genetic "potential".

It's the same way people gush about adorable animals. In my case, I'm a big fan of sloths because they're ridiculous and I'm stunned they've somehow managed to slowly outrun extinction for the last 40 million years. But from a practical perspective, what have sloths done for us lately? It's not like they're out there curing cancer. And if they are, they're doing it pretty fucking slowly.

How much time and effort do we waste talking about how adorable sloths are, when all they're doing is hanging out in the canopy eating leaves and occasionally falling out of trees? I would argue too much time, and I love sloths a *lot* more than I like kids.

Now I don't have anything specific against sloths *or* babies, but at least sloths are a lot easier to manage. Imagine you're driving your two juvenile sloths to sloth-school one day: they sit quietly in the back for a while, then they decide they're going to cause trouble by chasing each other around the back seat. Half an hour later you pull up at sloth-school to find they've barely climbed out of their seatbelts, and you hand them over to the zookeeper for the day.

But contrast that with a couple of kids. As you pull up at a set of traffic lights, you glance up at the bus next to you and make eye contact with a passenger. They look at you with pity and mouth "I am so sorry" as the screams from the backseat hit frequencies that will shatter glass. As the lights change and you pull away, the sounds die away. You glance in the rear vision mirror in fear as you suddenly realise it's now *too* quiet. You see your four-year-old has abruptly turned an unusual shade of green after eating a full bag of Skittles in less than a minute by pouring those multi-coloured sugar-balls straight down his throat. Before you have a chance even to ask where the Skittles came from, he looks at you like Peter Parker looks at Iron Man at the end of *Avengers: Infinity War* and whispers to you "Daddy, I don't feel so good".

A split second later, his entire body convulses as he tries to barf out his internal organs against the back of your headrest. You try to focus on driving as the half-chewed chunks of candy drip down the seat in a grey, amorphous mess, but you can't help flinching as the virulent stench of simple carbohydrates and stomach acid burns your nostrils like that first whiff of tear gas.

In the eye of the storm, there's a moment of complete and serene silence as the horror sinks in. You try to make a joke about the little guy "tasting the rainbow twice" because if you don't redirect your feelings into comedy, you'll drive this goddamn car off a bridge.

Confused silence reigns for another few seconds... before your six-year-old daughter in the other seat *just. starts. screaming*. A red-pink chunk of half-chewed Skittle slowly drips down her ballerina costume as she unleashes her banshee wail and waves her hands around like a penguin on meth. Meanwhile, her little brother stares wide-eyed in stunned silence - completely dumb-struck by the life-altering spiritual experience of power-yakking multi-coloured confectionary in a moving vehicle. And you? You're just trying to keep it together long enough to get home and look on the dark web for a "kids for sloths" exchange program.

There are so many folks with these cutesy sayings like "Kids are our only hope for the future" but I'm not relying on kids to improve the future: they're incompetent, poorly trained, and their fine-motor skills haven't fully developed. And just because they're physically smaller doesn't stop them from being even bigger assholes than adults either. A friend worked at a funeral home for several years and would see all the cards that kids would leave for relatives who'd passed away - one of the cards was scrawled in crayon saying thank you from the worms, while another just said: "I hope you have a nice time in heaven/hell".

Yet even knowing their potential to be tiny, amoral psychopaths, so many people see becoming a parent as the pinnacle of their adult life. Everyone coos about how newborns are "gorgeous" no matter how objectively ugly someone's crotch gremlin is.

Whenever I share my thoughts on breeding though, my friends are always quick to cry: "But Josh, you'd make a great Dad!". Which if you've gotten this far into the book, you'll hopefully agree is a *wild* misjudgement of my character. Unfortunately, the final entry in my treasured *101 Things to Do Before You Die* demands I try to "Continue Your Gene Pool", with threatening blank spaces to write the names of future spawn and room to glue in photos too.

Now while I'm committed never to be responsible for the development or nurture of anyone too young to remember VHS, I'm equally determined to complete as many of the *101 Things* entries as possible. So in 2016, I did the only reasonable thing left - I became a registered sperm donor, and I've been frankly shocked by the lack of support I've received from friends and family for it. When I told Mum she just rolled her eyes and said "Of course you did", while Dad laughed and called me a wanker... he was technically right, but it didn't feel necessary to say. Around the same time, one of my best friends unexpectedly emailed to tell me his girlfriend was pregnant and he was "going to be a Dad!". I excitedly replied: "Oh dude, that's amazing! I just became a sperm donor - we'll *both* be Dads!". He did not think this was funny.

Although If I'm honest, the results from my first visit to the sperm bank were frankly terrifying. Nothing scares someone determined not to have kids quite like a fertility specialist throwing around phrases like "Super Donor". Nope. Noooooope. Nope nope nope. Great news that donations in the future would be straightforward and that I could supply what they needed, but the overwhelming concern was now I'd trip walking down a busy street and someone driving a block away would be impregnated. Please don't for a second think that I'm bragging here either - it's just that having the potency of my spunk tested in a laboratory had created a genuine existential crisis.

While I've *always* been painfully careful about practising safe sex, previously my primary focus had been on not contracting super gonorrhea and assumed the chances of getting someone pregnant from a condom breaking or slipping were reasonably low. But thanks to cold and indifferent science, I *now* know I have more than double the average risk of catching the most dangerous sexually transmitted disease of all: children.

As soon as I heard the results, I asked the doctor about getting a vasectomy as soon as the donations were complete. But he just laughed and told me a healthy, single guy in his early 30's without kids would struggle to find a doctor who'd give him the snip because there was a "high probability" I'd change my mind and want to have kids in the future after all. And no - being shortlisted for a one-way mission to Mars *wasn't* enough justification for blocking someone's tadpole delivery system.

Not to be deterred, I did some digging and found I *could* get a safe and no-questions-asked vasectomy, but the fertility doctor had made another point on why I should hold off getting the snip that was far more convincing: call-backs. My five donations could start up to ten families, however just like a blood bank there were often shortages, so clinics would call up previous donors to see if they'd be up for another round.

Now my blood type is AB+, which means I have a full antigen set and can receive blood from almost anyone. But it also means I can only give directly to the same 2% of the population who also have AB+. I've also lived in areas with high rates of Malaria, Yellow Fever, Cholera, and Typhoid; and barely survived an acute form of blood poisoning thanks to Lyme disease.

So while for most people "every blood donation saves three lives", my blood would probably *take* three lives if I ever shared it. So it's just as well that donating semen is a *lot* more fun, and plenty of future parents seem to want that.

There was a mountain of tests to get through as they checked me for every possible genetic abnormality, and only after months of visits to the clinic was I deemed worthy of contaminating the gene pool. There was just one final hurdle before I could finally start the donations properly - two sessions with a psychologist talking about how I *felt* about cumming in a cup so other people could have kids.

We were less than two minutes into our first session before the whole "going one-way to Mars" thing came up, so any pre planned structure the psych might have had immediately went out the window - we were straight into how we'd get to Mars, what it'd be like to live there, and how we'd produce oxygen. All the usual "Holy shit, I'm talking to someone trying to live on another planet" kind of questions.

She had a few things related to sperm donation that needed checking off a list, like how would I feel if I met someone and wanted to have kids but didn't want my precious genetic material continuing to be freely available from a cryogenic vat? Well, that was easy - I'm going to Mars, so why would I have kids on Earth if I was then going to leave them behind? "Yes, Josh... but what if you changed your mind?".

Well, *not having kids* then changing my mind seems a lot easier than *having* kids then deciding I *didn't* want them after all - it's not like you can just shove them back in, and finding that sloths-for-kids exchange on the dark web again is going to be a real pain.

However, the weirdest turn in the entire experience was having the psychologist suggested I didn't want kids of my own because I don't want the "responsibility", and that fear of responsibility was cutting me off from one of the "greatest things a person can ever do". You know, because walking on another planet *isn't* the most significant thing I could ever do with my life - that's reserved for giving someone a creampie and then sticking around to keep the product as a pet.

On one level, she is entirely right though - I *didn't* and still don't ever want to be responsible for raising a bunch of whining shit-gibbons that look like me. But I do take offence to the suggestion I'm deliberately avoiding responsibility, because even from a climate change perspective the "responsible" thing is to not have kids at all. I don't mean it from a pearl-clutching "How could anyone raise children when the future is so grim!" kind of way. I mean the "having one less kid saves an average family nearly 60 tons of CO2-equivalent emissions per year" kind of way. Your kids aren't just stealing oxygen - they're also helping heat up the planet and speeding us towards the day the dolphins take over.

It's not a competition, but I feel like I'm taking on *greater* responsibility by pushing against the grain and *not* having kids. After eight years of speaking in schools, I hope we're all on the same page about how much I broadly dislike children. So why would I choose to have kids like the vast majority of people, instead of *living on another planet like no one has before*? It also doesn't have to be one or the other - some Mars One candidates have deliberately decided to have kids *because* they might be leaving the planet, and they want to help raise them before they go. Personally, I can't imagine having kids only to then leave them for a life on Mars, but that's just because I can't imagine having kids *at all*.

And while we're at it, let's acknowledge that I've I signed up to go *one-way* to another planet and live underground like a fucking wombat - I clearly don't have any issue with giving up my "freedom" and taking on the "responsibility" of an entire species when it doesn't involve reproducing.

After all this, I somehow managed to convince the psychologist I was sufficiently balanced enough to start donating my genetic material to Melbourne's cryogenic freezers, so the women of Victoria would soon be able to spawn tiny ginger rage monkeys. The psychologist also made sure I knew what would happen when my spawn were inevitably unleashed on the planet: whenever there was a live birth of a child conceived using my sperm, I'd get an email with the child's sex and birth month, but nothing else. Once the kid turned 16, they could then apply for my contact details and even meet if we mutually agreed to it.

Which is all well and good if you're planning to live out the rest of your life on Earth like a boring normie, but I suspect it'll be a little bit more difficult to "meet up" when I'm living more than 50 million kilometres away on another planet. Holy shit - could you imagine finding out on your 16th birthday that the dude who's sperm your parents picked is now living on *Mars*? "Mum, Dad? It says here the donor lives on *Mars*? There's no other address details or a postcode - it just says '*Mars*'. This is a joke, right? Josh Richards? You're saying the idiot who blew up one of the Mars rovers trying to super-charge it with alcohol he brewed in the greenhouse is my biological father?".

Naturally, I ticked every single box agreeing to my donor babies getting in touch - mostly because by the time they're old enough to reach out, I'll already be living on another planet.

They'll also be at least three light-minutes away, so no telephone or video calls - we'll have to pre record videos and send them back on a delay. And if they want to visit me on Mars, they'll need to be at least 18 to apply, require several years of training, and then spend another 6-8 months getting to the red planet itself. So no matter what, they'll at least be old enough for a quarter-life crisis when we meet face-to-face.

Legacy

Long after the entire process ended though, the psychologist's suggestion I was running from responsibility still stung. I turned it over and over in my head, trying to work out why my response to "avoiding responsibility" and "craving freedom" had been so fierce - had she touched a raw nerve? *Was* I claiming not to want kids because I was subconsciously avoiding responsibility? Or was she misinterpreting my true motivations? Eventually, I realised my reaction had nothing to do with "responsibility" or "freedom": it was because she'd suggested that having kids was a more important *legacy* than the one I was committed to creating.

For many people having kids *is* the most excellent adventure they'll ever have, and I genuinely respect that. However, it's also about creating people bound by blood to remember you. That desire to procreate blends with a desperate but impossible desire for immortality, and it can drive people to extreme lengths to cultivate their legacy. Initially, I signed up to be a donor purely for the laugh of trying to complete one of the *101 Things*. But there is no way I'd have committed to the endless blood tests and the sheer weirdness at jerking off in that weird little room if I hadn't one day realised the extraordinary lengths so many people take to have kids.

It was a simple poster in the fertility clinic's waiting room that forced me to instantly switch from "This will be something funny to talk about in a comedy show!" to "This is something I have a responsibility to do". While there were plenty of pamphlets covering the eye-watering costs of attempting artificial insemination, this poster displayed the mind-boggling number of medications and injections a woman has to go through while trying to conceive via IVF. Staring dumbstruck at it, I figured if people were willing to go through all of that just for a 50/50 chance of having a kid, then how much would they love and cherish that kid once it was born?

If becoming a parent was someone's lifelong goal - their chance to embark on a most excellent adventure by creating new life and a human legacy - then the least I could do was help provide some of the raw materials they'd need. There's no doubt they'd be a hell of a lot more committed parents than I ever would. While people pursuing IVF might auction their house to have a baby, if I had kids they'd probably inspire an episode of *CSI: Cyber* where the team bursts into someone's home to stop an online baby auction.

Strangely, it was the whole "kids as legacy" element of the psychologists' question that had truly upset me - that I'd go to another planet and regret not leaving a "legacy" on Earth. She figured I was trying to go to Mars to avoid raising kids and have people remember me as one of the first people to walk on another planet instead. I didn't understand it at the time, but I realise now I was furious with her suggestion because I'd committed to going to Mars years earlier for almost the exact *opposite* reason. Because if the name Josh Richards disappeared from all memory and history, that would be absolutely fine by me - I couldn't give a shit about my legacy. Anytime someone tries to provide me with even mild praise, I get awkward. So there's zero desire to see my name in lights or to be remembered for all-time.

What I *do* care about is helping people. I care about seeing our species work together to do something extraordinary. Care about seeing people embrace the best parts of who we are as a species and pursuing what some say is impossible. It's why I've spent the last eight years visiting schools and willingly speaking to over 100,000 of your truly awful kids - not because I "love inspiring the youth", but because I care about where we're heading as a species. People gush about finding happiness, but my "happiness" doesn't mean shit, especially when I'm doing meaningful things to help humanity become an interplanetary species. What we're trying to do here is so much bigger than me, than you, our family, friends, country, or anything else.

Hell, I'm not even fussed about leaving a *historical* legacy as the "first on Mars". Journalists regularly ask "What will your first words be on Mars?" and then look shocked when I say that no one except hardcore space nerds will ever remember my first words on Mars... because I can guarantee I will <u>*not*</u> be the first person to step on the red planet. If we're taking a crew of two men and two women, why the *fuck* would you get yet another generic white dude to be the first person to walk on another planet? The Apollo program was extraordinary for putting 12 sets of boots on the Moon, but every pair of those squishy white boots belonged to squishy American white guys.

With a gender-balanced crew selected from around the world, even if I *did* somehow manage to squeeze on to that first crew, there's no way someone who looks like me should be the first one to step out of the landing capsule onto the surface. Standard white boys do *not* need more astronaut role models, especially when women make up just 10% of all the people who have ever been to space.

Having the two women on the crew step out first is an opportunity to provide some desperately needed role models for the billions of kids who don't look like me. And that's the legacy I want - more kids trying to explore space no matter what their gender, nationality, or deep-seated misanthropic tendencies.

Of course, the flip side to all this lovely gender representation is if I'm the *third* person on Mars, then I can say *whatever the hell I want* when I step out! Does anyone know what the third man on the Moon said? Or even who the third man on the Moon was? Anyone *except* the hardcore space nerds?

Charles "Pete" Conrad had a bit of a reputation for being a fast-car driving, small-statured smartass. He was labelled "unsuitable for long-duration spaceflight" after quitting NASA's first astronaut selection by dumping his full enema bag on the desk of the commanding officer responsible for testing. Pete was selected for the *second* astronaut group though, and in 1965 he helped set a new spaceflight record by remaining in space with Gordon Cooper for nearly eight days aboard a spacecraft Pete regularly referred to as a 'flying garbage can'.

Pete Conrad was supposed to be one of the best pilots NASA had, so attitude aside it wasn't a surprise he was selected as the commander of Apollo 12 and became the third man on the Moon. It probably *was* a surprise to anyone who listened to the landing though, because he decided to use his first step on the Moon to make a crack about being short, saying: "Whoopie! Man, that may have been a small one for Neil, but that's a long one for me". So no, I don't aspire to be like Neil Armstrong - I want to be like Pete Conrad, so future historians have to include footnotes explaining why some ginger idiot wasted his first words on Mars with a joke about keeping the red planet red.

I don't need to imagine some inspiring first words on Mars, just as I don't need to have kids so someone remembers my name after I die. I don't even care about doing something to make my family or country proud - it's just nice to be involved in something bigger than me.

Our species isn't going to be better by remembering my name. But if this little ginger space grinch has the chance to be part of something big and beautiful that'll help people by leaving Earth behind, then you better point me to the launch pad.

Appendix

Mars Project – Do. Or Do Not. There Is No Try
Originally published on <u>www.themightyginge.com</u> (September 17, 2012)

As most of you are aware, I had my 10 year high school reunion in May. And a month ago today I turned 27. There's been a lot of smart ass comments since my birthday about the <u>27 Club</u>, and how this year I need to take care of myself because things could change quicker than I could handle them. I don't really mind the comments, but I have realised just how different I am from the people I grew up with. How different I am from the people I've spent most of the last 2 years around.

It's been a strange 6 weeks here in Gingeland. First I finished work in Stroud, packed everything I needed into a backpack, gave away everything else, and jumped on the train to Edinburgh for my fourth consecutive year at the fringe festival. This was the year that I was going to break through – I had an amazing show concept (even if I was still writing parts of Keith the Koala's show the day before my first fringe performance), I had my gig camera ready to make fringe video diaries, and I had my bike for racing around the streets of Edinburgh to shove my sociopathic koala alter-ego onto any stage I could find.

Doing my parents proud

A week later I'd cancelled the remaining 18 performances. A week after that (and three days before my birthday) my laptop was stolen from a crowded office where I was running a podcast for a major fringe publication – my only real side-project after I pulled the plug on Keith's show. I managed to get a 5-star review from "Cream of the Fringe" for Keith's show, but the show remained cancelled. Why I pulled the show seems irrelevant now – what's more important was how good it felt deciding to cancel a show I'd put all of my spare time into for 8 months.

After the fringe ended I headed south to Brighton – I had a lot of friends living here already or headed this way soon, and truth be told I didn't have or know anywhere better to be. I thought moving down here would be a chance to settle into an easy day job, develop my comedy and regularly perform in Brighton or up in London. I'd even try to get a job working in radio: practicing my performance skills in my "day job" then performing at night. And moving down here went fairly smoothly – a friend moving here needed a second driver to get her stuff from Glasgow, I needed somewhere to stay until I sorted a job & a flat.

I was only here 24 hours before I was on a train over to the Netherlands for International Redhead Day, which I'll write about in more detail soon. For now let's just say I met some wonderful people who I know I'll have plenty to do with in the future, and bumped into an old friend from the Perth stand-up scene who I hadn't seen in years shooting a comedy documentary for SBS alongside one of my all-time comedy heroes. The trip was an amazing success, and I felt a sense of purpose and camaraderie I've only felt a few times before.

Get enough gingers together and shit is always going to go down. Also, check out that sexy, ukulele-carrying bastard left of screen next to the dude in blue... BLUE STEEL BABY!

I got back to Brighton and started applying for jobs – aged-care, royal mail jobs, camping stores, anything. Then I waited to hear back from them. Which I'm sure you're all aware is something I don't do well. So rather than sitting in coffee shops, discussing the latest fashions in Milan, and generally being a prick like 95% of the people who populate the streets of Brighton during business hours; I hit the library to research my next comedy show. This would be a return to an "Apocalypse Meow" style show: researching a serious topic and presenting it in an informative but hilarious manner.

Because everyone knows any serious show about the apocalypse needs a reading of The Very Hungry Caterpillar

I'd been researching for about 10 minutes before I found it. And I read it, went and researched something else, and came back. I tried researching something else, and came back. Went and had a coffee, and came back. For the last week, I keep coming back to it. And every waking moment since (and a lot of my non-waking moments too) I keep coming back to it.

All through the fringe and in the weeks since, I've been re-assessing what I really want to be doing with my life. I thought it was comedy, and that's partly true. I used to think it was the military – turns out that's true in a way too. Before that, I was sure it was science. And through a series of bizarre coincidences – particularly over the last three weeks trying to decide where to live and what to do since the fringe finished – I found a solution that brings my experience in science, comedy & the military together to achieve something I've secretly dreamed of since I could talk.

The closest anything had ever come to consuming me like this was the way re-joining the military did in the months before leaving Perth for the UK to apply for the commandos. But the circumstances are radically different, I've grown up A LOT (regardless of what you may think of me and my koala alter-ego), and even joining the marines didn't permeate everything I do like this has. I also know even with the commandos I've never had a clear plan for my future that reached beyond a year, much less the 10-year plan this gives me.

So rather than drag this out any further, here's the story. On May 31st, "Mars One" – a privately held Dutch company – announced their plan to launch a rocket in September 2022 that would put people on Mars by 2023.

They intend to use currently available technology to send a manned mission to Mars, and they intend to fund it by broadcasting the Mars settler selection process through a "Big Brother" style media event. And I say "Mars Settler" for a very good reason – it's a one way trip. Like the Pilgrims leaving the Old World to settle the New, these folks are going to Mars to stay. They'll be the vanguard, with the 4-person team landing on Mars in 2023 to be followed every 2 years afterwards by another team of 4 intrepid souls.

These will be the people who take the first steps beyond our planet without planning to come "home" again. And I'm going to be one of them.

Each team brings a new living module

The comedy show I'd started researching was on the simple fact that if we want to send a manned mission to Mars, the return trip is the most expensive and technologically challenging aspect – so what if the people who went were willing to stay for good? The plan was to write a show about the challenges a Martian settler would face on the way there and once they'd arrived, then tell the audience why I am both willing and an excellent candidate to go – now I have a chance to prove just that.

I don't need to justify here why I'd be an excellent candidate for the Mission Engineer position in the initial 4-person settlement team – people who know me well know exactly why, and anyone who doesn't will see soon enough. But the simple fact is that in exactly 10 years a rocket is going to launch to another planet, and I want to be one of the first people to step out of it at the other end. I know why I want to do it, and I know I will.

Things are already falling into place on this, but it's going to take time. Rather than announce things I'm trying to achieve, I'm going to announce them as they happen. But I will say I doubt I'll be staying in Brighton, and if I have my way then my new friends from Redhead Day will be teaching me Dutch pretty soon.

Things may be a little quiet here initially, but rest assured this is happening. I guess all I can say now is "Watch this space"

Always end on a bad space pun, so they're confused as to whether you're serious…

Sarah's Letter

Letter written by Sarah Young in February 2015, after Josh's selection as one of 100 final Mars One astronaut candidates.

FOR JOSH IN TIMES OF DOUBT
Sarah Young: Letter Of Recommendation For Josh Richards

Dear Josh,
This is my letter of support in times of trouble and worry. I write to you under the influence of strong emotional angst and unconditional love and friendship. The purpose of this letter is to provide you with everything you need to know about yourself to achieve your dreams of becoming one of the first members of the Mars One Project.

Josh Richards, you are the greatest person I have ever known on this earth. Soon you will be the greatest person I have ever known on Mars. I know how hard you have worked during this long procedure. I know how many hours you've spent tirelessly working towards educating and involving our future generations *and* the general public in the Mars thing. And a *book!* Jesus. Pick a thing.

I know you believe in the mission, and I know that you want to do this despite the costs to you. I have never actually met a person who not only cared about his friends and colleagues but the entire human race. You realise that the world is bigger than you, and that life extends beyond what most people imagine (Blogging, Facebook Rants and ISIS). I know this is true when I hear you talk about it with me (and all our friends…) You've highly qualified, and your skill set and experience is like no other.

You are a strategist and the type of person who can remain focused in a crisis (not to mention that your experience in the military/[Actress Name Redacted] would make you an outstanding candidate for dealing with stress/bitch-faces).

But this mission needs more than an outstanding candidate. I'm not naive, Josh Richards, I know that there are many, many others who share the same and perhaps higher qualifications than you do. But none of those people will ever be the same candidate as Josh Richards. I don't believe there will be any other candidate who shares the same qualities as you.

Here are some:

- You are a problem solver who can think quickly and rationally under stress. I have seen you enthral a group of primary school children - *primary school children!* And they loved listening to you talk about the Mars One Project because *you* were so excited. Space is big too. That's why we call it "space".

- And that's another thing you might or might not be aware of: your capacity to learn. You have the most inquisitive brain that never tires. I'm sure it's exhausting sometimes but that is the problem with being a big think... so much "space" to store information.

- You never allow your stress or anger to rise unnecessarily - one of your best qualities. It's not that I don't think you possess anger - I have just never seen you take your anger out on others. Maybe in the past but I don't know. But that is an even better trait. If you have been an angry person in the past you're conduct now makes up for it.

- You have the capacity to grow and learn from mistakes. As a person who (regretfully) is a little bit guilty of losing her shit, I notice that you never do. I'm glad Keith is there to help you. And I'm glad Keith has prevented you from becoming a violent murderer. You use all your powers for good. I laugh many times at the thought of you chasing cadets around like a little red psychopath.

- I have only ever known you to put yourself before others, even in situations where your own life is at risk. You've always shown respect towards me and others - even those you don't like. You have a very kind and gentle nature and a good way with people. I know that in the stress of a cramped space-cabin, people are still going to welcome your presence.

If I was to choose a human being to share this Mars One Mission with, I would hope that human has at least a quarter of these traits.

I have never said goodbye to a friend who *left the fucking planet before*. That sentence makes me laugh. It sounds so ridiculous and far away from human understanding. I'm sure you're scared (pffft) of what that might feel like. And your friends and family (selfishly, I know) might not like or agree with your choice. Because we will miss you and we will worry about you. I love you, Josh, and it terrifies me that you might go to *another fucking planet*. I will miss talking to you, joking with you, winding you up and walking with you. You have been a calm, guiding voice to a sinking ship for me. And I don't know what it will be like to think of my friend ON MARS!!! I'm already terrible at computers and I don't want to imagine trying to Skype you from Mars.

But this interplanetary adventure is part of your destiny and you must go and do hero things now. Be prepared for the fear, the isolation, the tinge of sadness and the regret sometimes.

But being a hero means more than doing hero things on other planets, it's being unbreakable in the face of uncertainty, in the face of criticism and doubt. So you must go. Humanity needs someone like you. I don't know many people who are willing to sacrifice their friends, family, birthdays, beaches, forests, mountains, long walks and the view of the stars from a tiny, pretty blue planet, just to advance the human race that little bit further.

And if for any reason at all, you chicken out of this opportunity, here is the process of what I will do. I will dedicate a large portion of my life to hunting you. You will not even know I am there. You know the way a Great White Shark stalks Josh? It sinks its dorsal fin under the water so it won't be detected. If for some reason you pass this opportunity in the face of self-doubt or *any* other equally lame reason, here is how you will be murdered by me:

<u>Transcript for the Dateline Mystery Episode 99 "Mars or Death" that was made in the wake of your murder.</u>

Voice Over:
… But Richards had no earthly idea what he was about to face. A man who had trained for years in the military, a man who had dealt with threats to his life on more than one occasion, who often found himself staring into the face of death many times before, had no idea what was about to happen.

While he was in a bathroom of a local comedy club, the place where Richards had first found his love of stand up comedy, the killer was lurking. That killer, Sarah Young, was on a mission… but not a mission to Mars!

While Richards stood in front of the bathroom mirror the killer lunged from behind a door.

DR. Phil McGraw, psychologist: Well apparently, Sarah Young had waited in the men's bathroom stall for up to six hours. Just *waiting* for this opportunity. The investigators actually discovered that she had fastened a door hook onto the back of the bathroom door so that she could literally "hang out" there waiting.

Young grabbed Richards from behind suddenly. The surprise caused his common sense to "go spastic" and he was unable to fight back. His attacker then used a garrotte and began strangling Richards.

An open mic comedian who was present at this time, recounts his experience:

Open mic Comedian (witness):
It was my first time trying comedy for the first time… I was really nervous because all my family had come to see me… I was waiting for my turn to get on and that's when I heard it… Oh God… I heard the shrieking of… well I thought it was a weasel…

Upon hearing these terrifying shrieks and believing it wa a weasel, the comedians exit the building, unknowingly leaving Mr. Richards to his attacker…

Sarah Young, a close friend of Mr. Richards, had heard only hours before viciously attacking the victim, that he had "pussed out" of one of the most important endeavours of the human race. He had recently become doubtful of his own brilliance and had changed his mind about travelling to Mars, and resigned from the Mars One Project. Mr. Richards' failure to realise his greatness, cost him his life.

Coroner Joe: She had used a garrotte technique used by the CIA and we are still trying to confirm exactly *how* this woman managed to obtain this information… possibly from a Youtube video… all we know is that this attack was carefully prepared and expertly executed.

The victim's head was almost entirely removed. The shocking aftermath was discovered by an "open micer", who has now become so traumatised, that he tore off his own face. Young was heard to say "I warned him" as she was escorted by police from the scene before escaping with Mr. Richards head. Although Young's whereabouts are still unknown, the head of the victim was later found in the luggage of a Mars One Project crew member. Mr. Richards' head is now on Mars.

You heard a noise. You looked up to see me for a split second before your death… Your eyes will ask "Why Sarah?" and mine will flash "Because it's your destiny" in return.

Go to Mars.

Sincerely yours,
Sarah Young

Drawing - "Welcome To Mars"

Gift from 12 year old student after speaking at a primary school

Daily Habits

Habits & principles to live by while preparing for Mars One selection

---

Physical

Get up ridiculously early - Start before the world does (You have a job to do)
- Wim Hof, Meditation, Early Workout on an Empty Stomach, Cold Shower
- Food - After writing (~midday) Protein, fibre, healthy fats [Eggs, avocado]
- Coffee - 1 hour after waking
- Move Every Hour - Work in 1 hour blocks, Drink Water
- Screens Away >30mins before bed - Bed by 8pm (Journal/Read)
- Wash sheets weekly

Mental

Read & Write EVERY Day
- Journal 3pg warm up, Publish 1000 words
- List Have To & Want To
    - Sometimes both, erase from Have To
    - WANTS>>HAVES

Go Solo - Cultivate "fertile solitude"
- Read/Draw over TV/Social Media)

Emotional

Say "No" - The World keeps turning without you
- Fun, Knowledge, Money - Only say yes if two out of three apply
- 80/20 rule - 20% of people & activities provide most value

- Rule of thirds - Explaining is draining, If it's a struggle it's too complicated
- Listen to your body - "Fear" vs "Growth" & Swear Off Complaining
- Lighten up with nonsense - Dot, TF2 Demoman, Hitchhiker's Guide, Stubb

Vulnerability Will Set You Free - Be an open book & peaceful warrior
- Dare mighty things, Allow yourself to suck, fail spectacularly & say "Fuck It"
- Die doing something boring people tell their kids about as a "cautionary tale".
    - We're all going to die - I intend to deserve it
- Do the Unexpected - Howling Mad Murdock, Anarchist/Artist not a scientist
    - Thwart institutional cowardice (Ask forgiveness not permission)
    - Figure the work-around - Guerilla tactics are best
    - Who/What are your gatekeepers? There are no gates

Four Agreements
- Be Impeccable With Your Word
- Don't Take Anything Personally
- Don't Make Assumptions
- Always Do Your Best

<u>Spiritual</u>

Meditate
- Focus on Gratitude, Just say thank you, Act in highest interest of all
- Empower others - People are innately good, but often scared

Cut All Attachments
- Expect nothing, Embrace the ebb & flow of the universe
- Connect to the "perfection of creation" [Anders - BSG]
- This too shall pass - always evolving & accelerating change in others
- Only constant is change [Ouroboros, Nonlinear time in "Arrival"]

Laugh at Reality - It's not to be taken seriously, and there is comedy in everything
- There are no ordinary moments - "Humour. Change. Paradox"
- Albert Camus "Find your question", Douglas Adams "Can't have question AND answer"

Live Like You're Going To Die In A Year
- Stop "Busy", be "Interested"
- Love what you do & do what you love [Alan Watts] <u>You owe others nothing</u>
- Only Here & Now - Fears of the future & pain of past are useless
- Embrace Abundance over Scarcity

Use your intuition - Follow the Ginge
- All the answers are within you - listen to your little voice
- Act, don't react - You always sense the answers far ahead

What is your deep sense of purpose? What is your "why"?
Reminder: Every candidate wants to help humanity
What would you do if you were launching to Mars in a year?
A month? A week? A day?

Writing

Principles to guide writing, developed while writing about Mars One

Write EVERY Day - Don't die with your story untold
- Drink Coffee & Take a Huge Shit
- Read Every Day - You'll only get one or two ideas from even a good book
- Turn up for work - No discipline = no creative freedom.
- Trust creativity & love what you do without reward

Don't Ask Permission - Tell stories how you would to friends
- Be vulnerable & honest - Share something nobody knows about you
- Write in the same voice you speak with (read it out loud)
  - If it sounds boring, kill it
- Be scared of what people will think of you, but don't deliberately hurt anyone
- Punch upwards - Purge cynicism & half-smile through all
- Live weird and laugh - What makes ME laugh gleefully (Jungian Vs Freudian)

Simple writing - Use lots of periods and no semicolons
- Paint pictures with words - relate to people by making them laugh & cry
- Emotion cancels logic - Answer call of sea [Saint-Exupery]
- Heartfelt language over excessive adjectives and complicated nouns
- Nest smaller stories inside a grander one
- Use "said" instead of ANY other word

Let it sleep - Stretch, Coffee, Read, & Let Poor Work Go

Mars One - Astronaut Selection Process

Summary of Mars One's selection process including candidate numbers
https://www.mars-one.com/mission/mars-one-astronauts

Selection Round 1: Online application (202,586 to 1058)
- General candidate details, motivational letter, resume, 1 min video. Could initially apply in one of 11 most used languages on internet

Selection Round 2: Medical (1058 to 660) & Interview (660 to 100)
- All profiles made public (open about commitment)
- Medical: Similar to NASA & ESA
    - Good eyesight, general health, no drug dependency, full mobility and free joint movement)
- Interview: Study the risks & dangers of the mission
    - Radiation exposure, shielding required, reserve water & O2 in storage
    - Likelihood of being team players (place team ahead of self)
    - Final question to reveal real reason for being part (sincere about settling)

Selection Round 3: Group Challenges (Expect 100 to 40)
- Indoor & Outdoor, testing ability to work in a team within limited conditions, interdependency, trust, problem solving & creativity skills, thoroughness & precision, clarity & relevance of communication
- Candidate knowledge of study material (provided in advance) is essential to pass
- Candidates eliminated based on behaviour both inside & outside challenges

Round 4: Isolation (Expect 40 to 30) & MSSI (Expect 30 to 24)
- Isolation: Self select teams with greatest diversity (age, nationality, ethnicity, gender balance)
    - Selection committee setup group dynamic challenges & provide study material
    - Candidates will have to consider & prepare basic facts about self & personal preferences
        - Small things matter (leaving out dirty socks, dirty dishes, body odour, ect)
    - Study material related provided in isolation unit before final challenge
    - Multiple challenges (including study material tests) to select 30
- Mars Settler Suitability Interview (MSSI)
    - 4 hours and videoed
    - Questions on teamwork & group living skills, motivation, family issues, performance under stressful & unique working conditions, and judgement & decision making

Mars One - Five Key Characteristics

Critical astronaut characteristics required in Mars One candidates http://www.mars-one.com/faq/selection-and-preparation-of-the-astronauts/what-are-the-qualifications-to-apply

Resiliency
- Your thought processes are persistent
- You persevere and remain productive
- You see the connection between your internal and external self
- You are at your best when things are at their worst
- You have an indomitable spirit
- You understand the purpose of actions may not be clear in the moment, but there is good reason - you trust those who guide you
- You have a "Can do!" attitude

Adaptability
- You adapt to situations and individuals, while taking into account the context of the situation
- You know your boundaries, and how/when to extend them
- You are open and tolerant of ideas and approaches different to your own
- You draw from the unique nature of individual cultural backgrounds

Curiosity
- You ask questions to understand, not just to get answers
- You are transferring knowledge to others, not simply showcasing what you know or what others do not

Ability to Trust
- You trust in yourself and maintain trust in others
- Your trust is built on good judgment
- You have self-informed trust
- Your reflection on previous experiences helps to inform the exchange of trust

Creativity/Resourcefulness
- You are flexible in how an issue/problem/situation is approached
- You are not constrained by the way you were initially taught when seeking solutions
- Your humour is a creative resource, used appropriately as an emerging contextual response
- You have a good sense of play and spirit of playfulness
- You are aware of different forms of creativity

Josh's Notes - Five Characteristics

Personal notes on Mars One's five astronaut characteristics

Resilience
- Having a "Why?" - You can endure anything if you have a reason to
    - Meaning >> Happiness -"Happy" is a poor term [Perfectly Unhappy]
    - Attitude/mindset is everything [Always capable of more than you think]
    - Motivation is Overrated - No excuse not to finish a project
- Success is determined by how much pain you'll tolerate
    - SEALs "Only at 40%" - <u>You Can Dig Deeper</u>
- "Commando" self-sufficiency
    - Mindset - First to understand, to adapt, to respond, to overcome
    - Values
        - Excellence. Strive to do better
        - Integrity. Tell the truth
        - Self-Discipline. Resist the easy option
        - Humility. Respect rights, diversity and contribution of others
    - Spirit
        - Courage. Get out front and do what is right.
        - Determination. Never give up.
        - Unselfishness. Oppo first; Team second; Self last.
        - Cheerfulness in the face of adversity. Make humour the heart of morale.

<u>Adaptability</u> - Humans are more adaptable than we expect [first trains/spaceflight]
- Comfort Zones - Hardness Vs Toughness (Vulnerability)
    - Shackleton's Way - Focus on what's needed to survive
    - Expertise comes from where you've made mistakes.
        - Teach what you want to learn/master
        - Look where there is darkness - WHY does it scare you?
        - Go a little further [Bowie] & Never go back to the carpet store
- Nomad - Self-sufficient and adaptive
    - Practice basics every day - Make it stupidly simple
    - You perfect whatever you practice
    - Possessiveness: Walk away from ownership & attachment
    - What would the world be better without? You? (Arthur Dent)
    - Be a creator who needs nothing & has no attachments
    - Build what is needed when it's needed & give it away after
    - What can you remove? Purge Physical & Emotional
        - You can't own what you can't carry
        - Things you own end up owning you
        - Make a list of things to stop
        - Always clear to neutral
    - Rule of thirds: Celebrate +ve ⅓, Invite neutral ⅓, Cut -ve ⅓
        - Cut what drags - explaining is draining
        - Give what you're not using to someone who will

- MacGyver - Problem Solving
    - Be a "Do-er" (Learn by doing, not discussing)
    - Defined resources (Know what you have & how much time)
        - Clear goals & deadline (Know your stuff & use adrenaline)
        - Don't ask permission (Create an enabling environment)
- No perfect - Just about making choices
    - Doing vs. talk - Focus on the art of doing, not result
    - Don't plan/think/analyse/hope
        - "Do. Or do not. There is no try"
    - Don't complain about it - fix it, because no one is coming to save you

Curiosity - Problem Solving
- Purpose is singular - run toward it with calm & steady focus
    - Don't wobble - act, sleep, laugh. Misery lies between
    - Regular & small rather than erratic & big

Ability to Trust - Integrity
- Always be honest
- Your role as communicator/scribe - It's not about you
- There are no "teams" (all on same team), Lead by Example
- No "winning" but it is all a game - demonstrates how you interact

Creativity/Resourcefulness
- Make trouble doing the right thing [Red team]
    - Asymmetric Warfare = Disruption
    - Great teams need thought diversity
- First Principles Thinking (Vs Analogy)

Josh's Notes - Mars One's Book

Personal notes on the essays in Mars One's book "Mars One: Humanity's Next Great Adventure" published 2016

"Improvisation and Exploration" (Mason Peck)
- Most important technical skills is <u>ability to improvise</u>
- What makes spacecraft work today is human element
    - Problem solving, creativity, innovation
- Hacker ethos: Make do with what you have and be creative

"Medical Skills For An Interplanetary Trip" (Thais Russomano)
- After a 6 month ISS expedition NASA does rehab of 2 hours/day for 45 days
- Most valuable medical data is from Mars One colonists as guinea pigs
- Need to be in good shape immediately after landing (incremental task loading)
    - Exercise countermeasures are critical

"The Politics and Law of Settling Mars" (Narayan Prasad)
- Outer Space Treaty (1967)
    - Forbids any nation on Earth from claiming sovereignty over any celestial object
- Moon Agreement (1979) [Good Mars template]
    - Used exclusively for peaceful purposes without disrupting the environment
    - UN should be informed of location and purpose of any station/outpost
    - <u>Controversial:</u> Moon as common heritage of humankind, and international regime should be established to govern exploitation of resources

- United Nations Development Program (UNDP) could be good Mars template
- Law has usually lagged behind scientific & technological advances
- Lack of political interest in developing law → non-gov actors declaring own code of conduct

"Food for Mars: Cycling to Mars"
- Moon & Mars soil simulants → Seeds and potatoes
  - Rye, radish, garden cress & pea seed (Proving full life cycle) [Multiple plantings/germination cycles in Moon/Mars soil]
  - Better germination rates in Earth soil
- Researching heavy metal uptake (Zinc, Lead, Mercury & excess iron)
  - Introduce radish, pea, rocket, spinach, rye, tomato, green beans, carrot & potato

"Food for Mars: Wieger Wamelink Q&A"
- Biggest crop challenge is energy
- Sustainable crop growth in Martian soil requires Fungi, bacteria & pollinators
- Substrate & Water culture
- Underground chambers with LED lights
- Urine can be applied directly, faeces need to be sterilised (unwanted human bacteria)
- Bumble bees are better than bees (ease of transport, Queen can hibernate for 6 months), Chickens as a potential after insects

Image Credits

Leopard seal2.jpg - Image accessed through wikimedia
https://commons.wikimedia.org/wiki/File:Leopard_seal_2.jpg
Licensed under "Creative Commons Attribution-ShareAlike 2.0 Generic"
https://creativecommons.org/licenses/by-sa/2.0/legalcode

Pebbles the Southern Hairy-nosed Wombat.jpg - Image accessed through wikimedia
https://commons.wikimedia.org/wiki/File:Pebbles_the_Southern_Hairy-nosed_Wombat.jpg
Licensed under "Attribution-ShareAlike 4.0 International"
https://creativecommons.org/licenses/by-sa/4.0/legalcode

Milky Way Galaxy.jpg - Public domain Image accessed through wikimedia
https://commons.wikimedia.org/wiki/File:Milky_Way_Galaxy.jpg

Arecibo message.svg - Image accessed through wikimedia
https://commons.wikimedia.org/wiki/File:Arecibo_message.svg
Licensed under "Attribution-ShareAlike 3.0 Unported"
https://creativecommons.org/licenses/by-sa/3.0/legalcode
Image attributed to Arne Nordmann, modified from colour to greyscale

Acknowledgements

There's an African proverb that says it takes a village to raise a child. I don't know if that's true, because I have absolutely zero interest in raising kids. But I do know it takes a village to write a book, especially one as deeply personal as this.

In the eight years since Mars One provided a clear direction to my childhood passion for space, I've met so many extraordinary people who have all encouraged me to pursue a life on Mars in whatever weird and wacky way suits me best. There are simply too many incredible space nerds to name you all, but please know that if I've talked to you about living on Mars at some point in the last decade - from the fine folks at National Science Week to my friends from the Summer Space Programs in Adelaide and Haifa - please know it's partially your fault this book was unleashed on the people of Earth.

Niamh Shaw has been an inspiration, an accomplice, and an incredible friend since we first met over Skype on St Patrick's Day in 2014. She's the kind of passionate science communicator and talented artist I wish I could be, if only I weren't such a crowd-avoiding anarchist. I'm genuinely honoured to have her write this book's foreword, and I can't wait to see her achieve her dream to become the first artist-in-residence on the International Space Station.

Writing *Cosmic Nomad* has been a drawn out process that started in late 2015, and without the on-going support of my fans on Patreon this book would never have passed the "nice idea" stage. Thank you so much to Dan Middleton, Fay Wells, Jessica Hausauer, and all my supporters for contributing through Patreon so I can keep writing.

Most of the ideas in this book were swirling around my head as disordered chaos until early 2018, when I had nearly five weeks of quiet solitude in a wooden hut on Flinders Island. So thank you to Mountain Seas retreat for the artist residency I so desperately needed, and to Priya Kitchener for feeding me when I forgot to eat and for being a regular source of mischief when I took my writing (and myself) too seriously.

The art of writing books is supposed to be in *re*-writing them, and Georgi McLaren has had to share a roof with my grumpy wombat persona through that challenging "re-writing" phase. Thank you for your patience while I've locked you out of the office, stomped through the house each hour to "get my steps", demanded you read paragraphs and then pulled faces when you've politely asked if we could spend time together instead of me continuing to stare at a laptop screen. Likewise, thank you to Chloe Reid for regularly distracting me from editing, knowing full-well the moment I put the laptop away all I would do was babble constantly about book progress, cave diving, and rebreathers. This book would have been impossible to complete without the amazing support I've had from both of you, so thank you for being loving parts of my life and for tolerating my weird wombat ways.

Thank you once again to Anna Piper Scott for her incredible work designing *both* of my book covers, and the support and notes I've had from my proof-readers has been invaluable too - huge thanks to Carly "You Do You" Bodnar, Lisa "Rabbit" Stojanovski, and Shelley "Mum" Richards for all you've done. You're all incredible women, you've all played different but important parts in my life, and I'm lucky each of you has been willing to help give this book some polish.

There's also a few folks I want to mention who have made a huge personal impact at different points over the last eight years, but I haven't seen in person for a while through circumstance or mutual decision. Thank you to Cameron Davis for being my inspiration in 2012 to leave standup for writing, and to Megan Leggo for housing a ginger hobbit in her garage as he adjusted to life as a future Martian and wannabe author. To Eli who was there at the beginning, to Lauren who forever changed how I see relationships, and to Amanda who picked up the broken pieces with some help from the monkey-dog Hugo - I'll always be grateful for the important roles you each played in this story, and I hope the parts of this book that sound familiar brought a smile and a laugh rather than any pain.

And finally, I want to thank my fellow Mars One candidates for being a constant source of inspiration. Dianne McGrath deserves singling out for specific thanks though - she's one of the most extraordinary human beings I've ever been lucky enough to meet, and I'm honoured to call someone so dedicated, empathetic, and kind my friend. Dianne has been an enormous support through all sorts of personal trials since we were both shortlisted in 2015, and it's my genuine hope that she walks on Mars before I do.

I've done my best to meet as many candidates in-person as possible, and every time I've been blown-away by Mars One's incredible talent pool. A very special thanks as well to Nat, Adriana, Kay, Hannah, Ryan, Robbie & Yari, Sara, PDP, Dan, Leila, Hampton, Megan, and so many others who have gone to such incredible efforts to meet up whenever we've been in the same country. I'm not really one for "tribes" but it's an honour to be part of this one. I have no idea what Mars One's future holds, but I do know that I'll see many (if not all) of you extraordinary people on Mars in the years to come.

Josh Richards is one of 100 global candidates for the Mars One Project - a mission to establish a permanent human presence on Mars by launching astronauts to the red planet one-way. After spells as a combat engineer, naval diver, commando, physicist, blasting specialist, fine art technician, and stand-up comedian, Josh finally turned his attention to his childhood dream of going to space.

Since 2012, he's spoken in hundreds of schools, universities, and businesses about how humanity will become a multi-planetary species, and how exploring other planets will improve life for those who remain on Earth. He's also the author of *Becoming Martian* - a humorous look at how colonising Mars will change humans in body, mind and soul.

He's currently based in Melbourne but looks forward to moving to Mt Gambier to pursue *another* life-threatening passion: cave diving.

www.ingramcontent.com/pod-product-compliance
Lightning Source LLC
Chambersburg PA
CBHW020320010526
44107CB00054B/1919